P9-DHN-669

SUSTAINABLE PLANET

SUSTAINABLE PLANET

Solutions for the Twenty-first Century

Edited by
Juliet B. Schor and Betsy Taylor
Center for a New American Dream

BEACON PRESS   BOSTON

Beacon Press
25 Beacon Street
Boston, Massachusetts
02108-2892
www.beacon.org

Beacon Press books are
published under the auspices
of the Unitarian Universalist
Association of Congregations.

© 2002 by The Center for a
New American Dream

All rights reserved

Printed in the United States
of America

06 05 04 03 02   8 7 6 5 4 3 2 1

Text and cover printed on paper
that is 100 percent postcon-
sumer waste and (text only)
processed chlorine-free

Text design by George Restrepo

Composition by Wilsted & Taylor
Publishing Services

Library of Congress Cataloging-
in-Publication Data

Sustainable planet : solutions for
the twenty-first century / edited
by Juliet B. Schor and Betsy Taylor.
    p. cm.
  ISBN 0-8070-0455-3 (pbk. : alk.
paper)
    1. Lifestyles.   2. Alternative
lifestyles.   3. Environmentalism.
4. Social justice.   5. Sustainable
development.   I. Schor, Juliet.
II. Taylor, B. S. (Betsy S.)
  HQ2042 .S88 2002
  306—dc21   2002011462

The chapters within this book are lovingly dedicated to the special children and grandchildren in our lives:

Augustus Taylor May and
Emily Anne Taylor May

Krishnan Dasaratha and
Elana Sulakshana

Jesse Broad-Cavanagh

Annie, Nora, and Christopher
Considine

Anna, Will, and Isabel Daly

Clara Dartt and Cole Jackson
Dittmar

Lucy Cordes Engelman

Willoughby Forbes

Jonas, Nora, and Stella Griefhan

Meika, Alexander, and Chiara
Hollender

Drew and Ava McDonough

Sophie McKibben

Kate Pipher

Eli, Daniel, and Rahmiel
Rechtschaffen

Jennie and Joanna Robin

Eva Rachel Spiegel

Margaret and Sarah Stratton

Gordie, Alice, and
Johnny Verhovek

Andre Wackernagel

The Children of New York's Twelfth Congressional District

And in loving memory of Rachel Anne Gaschott Ritchie, who left us far too early in life but whose spirit lives on

CONTENTS

PREFACE

The natural and social fabric of our planet is being ripped apart. As many as sixty thousand plant species, or one quarter of the world's total, are in danger of being lost by 2025. More than half the world's fisheries are already overfished, depleted, or in serious trouble. Water scarcity is increasing, and at an accelerating rate. The United Nations panel of over two thousand scientists warns that the planet is heating up rapidly—by as much as eleven degrees Fahrenheit this century. Nearly half the world's people, some three billion individuals (a number that has risen as globalization has proceeded), now live on less than two dollars a day.

How can we respond to these developments? What solutions will result in a better world at the end of this century? The view espoused by most political and corporate leaders is that environmental and social progress will result from *more* economic growth, the globalization of trade and commerce, and unfettered enterprise. They claim the ability to lift people out of poverty, protect the natural world, and generate more wealth through expansion of the global marketplace. This perspective assumes that the planet will provide infinite resources and absorb infinite waste. It discounts the need to set any limits on growth, despite rising human population. And it relies on an almost superstitious faith that the market will cure our problems. Of course, within this camp are those who recognize the need to better distribute the benefits of the global economy, as well as the need to restrain its more negative environmental aspects. This progressive wing calls for stronger democratic institutions that can set limits on corporate excess. But they share the perspective that the road to a better future is through greater consumption and economic growth.

An alternative worldview is taking shape. It is championed by scientists, environmentalists, relief workers, and human-development experts. They warn that we cannot continue to destroy the planetary ecology in pursuit of growth. They cite strong evidence that the market is failing to correct our environmental and human calamities.

Not surprisingly, proponents of the first view are dealing harshly with these challenges. Critics of globalization are ridiculed as "members of a flat earth society," despite growing evidence of the worldwide proliferation of poverty and despair. The dismissal of the scientific evidence is similarly stark. Early in 2002, Robert Watson, chief scientist with the World Bank and former Chairman of the International Panel on Climate Change, was attacked and removed from his position after he insisted that governments and corporations needed to reverse global greenhouse warming. The pressure to oust him came from the U.S. government, backed by energy corporations. Watson's views were simply too threatening for the dominant power structure—a network of political, financial, and corporate leaders who prefer to burn fossil fuel and maximize economic expansion at any cost. But Watson isn't alone. Scientists from many universities, think tanks, NASA, the National Academy of Science, and the Association for the Advancement of Science are warning of planetary catastrophes that stem, in part at least, from unfettered economic expansion. Yet for the most part, these scientific prophets are ignored. If they are even half right, the system needs radical transformation, from the household level to the global economy. For now, the power brokers cannot accept this and choose denial over truth.

Most of the authors in this book have concluded that we must cast off the nineteenth-century mentality that structures the current system. Along with thousands of other scientists, environmentalists, development workers, and social activists, they are proposing innovative models of sustainability that neither overstep the limits of nature nor accept current levels of global human suffering. Most support the positive elements of the prevailing order, such as technological innovation and enterprise that generates jobs and material wealth in order to ensure human welfare. But they also see the need for a dramatic change in course if we hope to conserve resources for future generations, curb global warming, and address human needs.

The new way of thinking is founded on the recognition of the interdependence of all life. It cherishes variety and beauty in the human, animal, and plant world. It is committed to the conservation of resources for future generations. And it makes a strong distinction between economic progress and economic growth. Proponents of the new models insist that the colossal scale of poverty and the growing disparity between rich and poor are not tolerable. And they blend a fierce commitment to economic and political democracy with a love of beauty, elegant design, and the wildness of nature. They see the emergence of an alternative in farmers' markets, worker cooperatives, healthy communities, land-use planning, socially responsible businesses, organic cotton, hybrid electric vehicles, barter networks, micro-enterprise, flexible work arrangements, simple living, reduced television watching, environmentally certified wood and fish, and a cultural renaissance of poetry, storytelling, dance, and reconnection to wild places. The new system is being built from the local level up. It celebrates public servants—teachers, firefighters, librarians, and health-care providers—rather than Hollywood and Silicon Valley celebrities. Its heroes embrace the values of sufficiency and ask first, "What is life for?" rather than, "How can I get ahead?"

Virtually everyone, including us, agrees that innovation and technology are essential. However, the evidence shows that the pressures from growth continually outpace the gains generated by increased efficiency and new technologies. Impressive improvements in heating, lighting, and appliance efficiency have been offset by palatial homes, big-screen televisions, and jumbo refrigerators, as well as increases in the number of appliances and electronic gadgets per household. If unsustainable consumption is a root cause of many environmental problems, we must fashion development strategies that generate income, employment, and security without increasing the throughput of energy, water, and materials. We must squarely face the question, "How much is enough?" But in so doing, we believe that we are no longer doomed to the sterile debates that have dominated the discourse on economics and the environment. Rather, the visionaries who have contributed to this volume provide a road map to an alternative way of living and being that is both sustainable and ultimately deeply satisfying. We invite you to share their journey.

INTRODUCTION

Bill McKibben

Pour yourself a glass of water. From the tap. Take a nice big drink, roll it around in your mouth like a connoisseur tasting some grand vintage. You've just done the one thing all animals that share the land or the air must do, the one bottom-line requirement for the maintenance of cellular life. Butterflies collecting at muddy puddles. Wildebeest trekking to waterholes. Coyotes slipping down to creeks at dusk. Don't dry out!

Now consider this. Recently, an internal Coca-Cola company website leaked into public view, detailing a new effort of the company, in conjunction with the Olive Garden restaurant chain, an effort it called H_2No. As in, drink Coke instead. "Research was conducted to better understand why tap water consumption is so prevalent and why consumers are making this beverage choice." Happily, the social scientists concluded, "It is possible to make other beverage choices more relevant to consumers." The "conversion strategies" included not giving people water when they sat down, and if they asked for it, informing them instead of "value offers like free refills that positively influence a beverage decision in favor of a soft drink."

In fact, the strong efforts of Olive Garden staff (who competed to win luxe vacations) managed to "reduce tap water incidence" dramatically. Not only that, by offering more choices, they claimed to be improving the lives of their patrons. In fact, said Coke, "The Olive Garden has sent a powerful message to the entire restaurant industry—less water and more beverage choice mean happier customers."

Here, perhaps, is the proper place to begin thinking about our lives as consumers—about what it means to live in the most advanced consumer society that there ever was, a culture that fills almost every cranny of our lives with a message, a product, an image. A place where everything but the air we breathe comes branded and hyped. (The water that we drink, certainly. Coke, for instance, now sells a water product called Dasani, a computer generated name picked after consumer testing showed it to be "relaxing and suggestive of pureness and replenishment." It is, in fact, whatever tap water is available locally—the bottler "enhances" it with minerals and subjects it to a process called "reverse osmosis" before pouring it into a blue-tinted vessel.) This book is an attempt to figure out what that society is doing to our economies, our planet, our families.

And to our souls—to our sense of who we are and why we are. Those senses drive our behavior, and so billions upon untold billions have been invested in understanding and appealing to our psyches. Most people now believe that the advertiser's understanding of our species is in fact a reasonable portrayal of "human nature"—that at bottom we are grasping machines, ever eager for more, supremely concerned with status, highly competitive. That we are, at bottom, *consumers.*

And this, in fact, is true. In every human brain I've ever come across, preeminently my own, there is a rather large chamber reserved for precisely this kind of self-absorption. Who knows what in our evolutionary saga drew the blueprints for this corner of our minds, but it is there. No use denying it—wise people never have. Indeed, most of our religious and ethical systems are attempts to deal with its reality, to tame and to limit it. What is different about a consumer society, though, is its celebration of this grasping quality, its insistence that this is the sum total of us. And its rebellion at the idea that we might limit ourselves in any way. Limits are anathema to homo-economicus—they threaten to slow down the constantly accelerating machine that we call our economy, a machine that can seemingly do anything except brake.

Another way of putting all this: Some years ago, in the process of writing a book, I watched every minute of television that streamed across the world's largest cable TV system in a single day: 2,400 hours of videotape, from country music videos to Abdomenizer infomercials, from the nightly news to the televangelists. The experiment yielded a fair number of useful insights, but the central idea that flowed through that coaxial cable, and that flows also through the shopping mall and the suburb and the theme park and the other crystallizations of our consumer culture, was this: You are the most important thing on earth. That is the central organizing principle of our age.

One would think that from such a starting place it would be only a short leap to happiness, to fulfillment. And indeed the consumer society has given us much: slightly longer lives (most of the big gains came from public health and sanitation laws); endlessly greater diversity of products, from food to cars; an unprecedented supply of entertainment; the freedom to travel great distances quickly, obliterating the constraints of time. We like these things—these things are woven into the fabric of our lives. And yet do they make us demonstrably happy?

That's an interesting question, rarely asked by economists, who assume the sum total of our purchases must inevitably equal contentment. Once a year since the end of World War II, however, pollsters have inquired of Americans whether they were happy or not. And the results are interesting: The number of people who described themselves as very satisfied peaked sometime in the mid-1950s. Since then, even as our material abundance has expanded exuberantly (one car became three, a three-channel TV in the living room morphed into a monitor in every corner of the house), our spirits have slowly but steadily flagged. It's as if we'd conducted a large-scale experiment to answer the question: Do riches bring happiness? Whatever else the economy can claim, widespread fulfillment is not on the list.

How could that be? Perhaps because that grasping part of ourselves is not "human nature" at all, or at least not all of human nature. There are other corners of our brain as well, other parts of our nature. Evolution has left us with quite a lovely collection of limbs and senses, muscles and emotions, intelligences and passions—we were not designed for lying on the couch and clicking the remote alone. And not designed for ourselves alone. At other times, in other places, other things have been at the center of our lives—the tribe or the community, nature, God. These central organizing principles yielded other worlds. Worlds that doubtless had problems all their own, and that we don't want or need to return to—but worlds that might give us real clues about how to rebuild our own society so that it worked a little better. Clues about how to rebuild our lives so they might fit us a little truer.

Those other worlds are a little hard to remember, caught as we are in the self-reinforcing marketplace ethos. Most critiques of consumerism have been pretty quickly co-opted. In the 1950s, for instance, there were plenty of people talking about conformity, about soullessness. But their prescription was usually more personal liberation, a suggestion that played into the great strength of the marketers, with their eternal emphasis on You. That is to say, "the sixties" as an idea was being sold persuasively before the decade was two-thirds over, a process that continues to this day, as boomers buy cars to the strains of "Revolution." If the focus on You has yet to make you happy, the logic of the consumer machine is simply to get something more. Surely it will do the trick—and if not it, then the next thing. Surely.

Think of it as an enchantment, a long, sweet incantation. If someone struck a discordant note, it was quickly incorporated into the lulling chant. The first real glass of cold water (or beverage of your choice, with free refills) poured over heads came from environmentalists, who had a few hard and straightforward questions to ask. Not about happiness, but about burning rivers, smoggy skies, disappearing landscapes. Did this, perhaps, have something to do with our "ways of life"? Were we consuming the planet?

Their queries, too, were deflected, at least in part. Engineers stuck filters on smokestacks and sewer pipes and insisted we could have it all, an unthreatening answer that appealed to many, since it put off the difficulty of thinking up some new way to organize our lives. But there were harder environmental questions to come. Take, for instance, global warming—it's less a result of some technical malfunction than simply of the sheer volume of our numbers and appetites. When we burn fossil fuel, we inevitably release carbon dioxide—there's no filter to put on the exhaust pipe. And we're now burning enough of it that scientists believe the planet's temperature will rise five degrees this century. If true, we'll live on a new earth, and not an improved one. The only ways to prevent it? Convert to new fuels like wind and solar, and use less energy—change technologies and change lifestyles. And do it fast.

If you want to understand what it means to be a consumer species, here is a number for you. Humans now consume 40 percent of the planet's photosynthetic productivity. Of the sunlight that falls on land and creates plant life, we take 40 percent. Barely half is left for the rest of all creation. Here's another number: By some accounts, human beings have used more material resources since the end of World War II than in all the time before. And so much of it has just been . . . well . . . silly.

Consider the sport utility vehicle. SUVs barely existed in 1990; by 2000, along with pickups, they comprised half the vehicles sold in the United States. Not because people suddenly needed to drive on rutted dirt roads—95 percent of them never ventured off pavement, which was just as well, since they showed an alarming propensity to tip over. Simply because the marketers had been at work, spending billions to spread the notion that You would be more manly (or, depending on your gender, safer) if you drove a Rover or a Blazer or a Navigator. And that, in any event, you would be more mystically in touch with the natural world. (Which in point of fact you were busy destroying—converting from an average car to an SUV and driving it a single year was the equivalent, in energy terms, of opening your refrigerator door and leaving it open for six years.)

The environmental attack opened the door, and other critiques followed. For instance, as people spent more and more time consuming entertainment, mostly from a feet-up position, Americans began literally to expand. Soon doctors were warning that, save perhaps for cigarettes, the greatest threat to public health was an epidemic of obesity. Overweight was no longer an aberration but a norm, as we converted to a diet consisting largely of foods that spent more on advertising than on ingredients (or, to put it another way, as we stoutly reduced tap water incidence). Another set of marketers was busy making us feel bad about the pounds their brethren had insisted we put on—soon "eating disorders" and "poor body image" were parts of a normal vocabulary.

Others began to worry about the effect of our way of life on our sense of community. In the small rural town where I've long lived, the older people can remember the advent of television—the way that "visiting" suddenly dropped way down the scale of interesting activities. It's hard to knock on the door when you see the blue glow coming through the screen, and harder still to carry on a conversation with someone who is glancing between you and the tube like a spectator at Wimbledon. But who really needs neighbors in a consumer society anyway? If there was something you couldn't do yourself (and there were more and more such things), you simply purchased what you needed. A man with a fiddle who used to play for dances was no longer in such demand because there was a phonograph/8-track/CD/Net appliance sitting in the corner.

Robert Putnam, the Harvard researcher who documented the decline in civic involvement in his book *Bowling Alone,* searched far and wide for the reasons: More women working? Less time? By far the strongest correlation was with increased TV-viewing—with the embrace of the central machine of the consumer age, the one that keeps whispering sweet nothings to You.

Oh, you could write a longer list than this. Violence, misogyny, depression—all seem to have some real and deep connection with this I-dolatrous way of life. But it is enough to say that nearly everyone has at least a few doubts about what our culture is up to. Survey after survey shows parents, for instance, believe their kids have become too materialistic—though perhaps the difference in the generations is not as great as some of those parents might wish to believe. Some of the shine has worn off the consumer machine; doubts are muttered about the true faith.

Still, as any economist would tell you, we *choose* to behave this way. We like this stuff. It's how we want to live our lives. Or is it?

Reading a business magazine recently, I came across a fascinating article. It was about "smart" homes, those "homes of the future," with remote-control everything that have been a staple of world's fairs for two generations. This one was "web-enabled," meaning that when Jared Headley, director of "consumer solutions" for Cisco Systems, taps on his computer tablet, "the window blinds in the room begin to lower, the overhead lights dim, and the TV set cuts out, as light jazz begins to emanate from grilles in the ceiling." "It's a date machine!" he says with a grin, going on to describe his wired dishwasher, his E-mailing fridge, his wired picture frame with updated images of the kids.

The interesting part of the story, though, is that it's not Amana or Proctor-Silex that is sponsoring the new work—it's companies like Cisco, Intel, 3Com, IBM, and Sun. "Though these products are accompanied by hype about saving consumers' time, giving them peace of mind, and even bringing families and communities closer together, the changing high-tech market is the main reason these companies are invading the home." What happened was that big chip makers and server manufacturers noticed that corporate customers were not upgrading their computers as often as before, leading to flat sales. In the words of one industry analyst, "These companies want to look at new territories to conquer. And that's the consumer market."

So, for instance, "By Net-enabling everything in the home, Sun hopes to create ever more Internet traffic. And more Net traffic means more servers—Sun's bread-and-butter products—purchased by broadband service providers. That's definitely the sweet spot for us."

The magazine points out the one weakness in the whole scheme: "Do consumers want all this interconnectivity? Will we ever really need our fridge to E-mail us that our milk is past its prime?" And the answer, of course, is no. No human being not suffering from some profound disability needs a machine to lower the window shades, turn off the TV, and turn on the light jazz. When IBM ran an ad showing a dishwasher that automatically phoned the repairman when it thought it was broken, consumers in focus groups rebelled—they didn't want the appliances making calls.

Nonetheless, the "service providers" seem confident that they can overcome consumer resistance, reducing the incidence of manual light dimming as effectively as the incidence of tap water drinking. "We'll sell a security solution, an entertainment solution, a messaging solution," says Michael Moone, VP of consumer business at Cisco. "There isn't a CIO or CTO of the home, so we have to give this to people in terms of things they already want and are familiar with."

And perhaps—quite likely—they will succeed. Not because we want or need these things; clearly we don't. But time and again we have bought into new technologies, new toys. How is this possible? Part of it is the skill of advertisers, who will surely figure out a way to link web-enabled homes to the love of our children, our fears for our security, and so forth. But part of it, too, is the simple volume of advertisement and hype in our lives, which in some deep way may make it harder for us to think clearly about what we actually do want.

Henry David Thoreau, in *Walden,* of course sounded the first real alarms about this coming culture. Not that nineteenth-century Concord featured the Internet or the satellite TV. Its main communications medium, in fact, was the wooden sign, one of which hung above each store and tavern. We would regard them as incredibly quaint—Thoreau, who had sensitive antennae, avoided the main street of town because they seemed incredibly intrusive, designed to "catch him by the appetite, as the tavern and victualling cellar; or by the fancy, as the dry-goods store and the jewellers . . ."

And he was right, I think. In a world like the one we live in, it becomes hard to hear your own desires. If you feel yourself wanting something, is that *you,* or is it the effect of a million commercials? For the sake of argument, use the following metaphor: Each of us has a small broadcast coming from our souls, telling us who we really are, what we really need. It plays constantly, but at an incredibly low volume. So low that it is jammed with ease by the static and noise of our information era. We do not know what we want—we are cut off from ourselves. Quiet, solitude, dark—all the conditions that might make it easier to hear that broadcast—are the scarcest commodities in this time and place.

We are frazzled, time-starved, isolated from our neighbors (three-quarters of us don't know those who live next door!). Those very conditions make us all the more vulnerable to the next pitch. A wired house? It will save time! (Though of course it won't—the time spent turning down the lights will be spent searching for the remote.) I can see the ads now: our homes as islands of calm, filled with light jazz emanating from the ceiling grilles (and nothing of the hours spent at work to pay for this vision, or the electricity sucked from coal and oil to make it all work). Were our feet firmly planted on the ground, such hype couldn't succeed. In fact, it took centuries to convince Americans to consume frivolously: The Puritan values of frugality and thrift were so strong that frippery and fashion were scorned. But all that has changed.

Or maybe, just maybe, not quite.

In the last few years, in place after place, small numbers of people have begun to try and fight back against the hyper-consumerism all around us. It is a hard job—less like battling air pollution than battling air—but activists, usually working outside the big environmental or social justice groups, have made their mark. Consider, for instance, the Center for a New American Dream, which has emerged as the central clearinghouse of a nascent movement, its website (www.newdream.org) the hub of many actions. But it is not alone—other groups and movements, working in an increasingly easy collaboration, suddenly abound. Just for example:

- TV Turnoff Week, after a decade of hard work, now reaches about five million schoolchildren annually. They and their families agree to turn off the tube, and at least for seven days get some sense of the unelectronic world, the one we used to call "real." Organizers have won the endorsement of a wide variety of groups—not just environmentalists and antiviolence campaigners, but also religious conservatives. Even more importantly, they've persuaded pediatricians—worried among other things about neural development and the link between TV and obesity—to issue guidelines urging that parents not expose their kids to TV at all before the age of two and limit it closely thereafter.

- Adbusters, a slightly anarchic Canadian collective dedicated to "uncooling" major brands, has made International Buy Nothing Day (that's the day after Thanksgiving, otherwise known as the start of the holiday shopping season) into a widely celebrated affair. Their mock billboards have spread far and wide, and they've helped spearhead the campaign to stop Channel One, the TV news-and-commercial network supplied to school homerooms across the continent.

• Co-housing, the European movement to build communities instead of detached homes, where families share meals and child care, is spreading even in individualistic North America. Community-supported agriculture—where folks pay a sum to farmers in the spring and then share in their harvest all year—has grown even faster.

• Churches, synagogues, and mosques are increasingly entering the debate. Heirs to traditions that emphasize God not mammon, they are starting to emerge as sources of potentially powerful critiques. For instance, a small but growing number of Christians are now celebrating alternative Christmases. In one such effort, which some of us christened "Hundred Dollar Holidays," pastors urge their flocks to spend just that sum—about a tenth of the American average—on gifts, and instead to look for joy in other, deeper rituals. A growing anti-SUV movement has drawn wide support from other Christians worried about stewardship of the earth: At one demonstration outside a row of dealerships near Boston, we held signs reading "What Would Jesus Drive?," an image that appeared in newspapers and on TV screens across the country.

This list could stretch on. Bike activists have organized Critical Mass rides in major cities to demand a share of the road. Others have targeted junk mail, or taken on the soda companies when they try to turn public schools into exclusive domains (reducing fountain-drinking incidence), or set up car-sharing pools. None of it has yet dented the statistics that really count in our society: Wal-Mart sells more each season, credit-card debt keeps rising. But it is hard to escape the sense that a new critique is emerging, not confined to (or even really noticed in) academic or intellectual journals but instead acted out in a hundred earnest and sometimes even good-humored ways. Those who have tried to label it have called it "voluntary simplicity"—the giving up of some potential wealth, some possible stuff, in return for time or fulfillment. In the words of author Vicki Robin, "Your money—or your life."

But it remains an inchoate world view, only occasionally swimming into focus. A few years ago, in Seattle, protesters descended on a meeting of the World Trade Organization. There was a certain amount of chaos, a few ugly anarchists, but mostly there were tens of thousands of people who peacefully took to the streets to proclaim a simple message: There are other bottom lines than the ones we've been told about. It's not just the economy, stupid, however much our leaders may insist otherwise.

Floating overhead the festivities, a large balloon carried the slogan: "Wake Up, Muggles." For any Harry Potter fan, no translation is required. But for those who haven't read them yet, Muggles are pretty much like us. They're the people who watch the TV, who run to the mall—and who somehow don't notice the loopy, wonderful world of wizards and dragons and general magic that exists all around them. A Muggle would be a perfect target for the H_2No campaign; a Muggle might even install remote-control operated light jazz-emanating stereo grilles in his ceiling.

But if the authors of this book are to be believed, we're not doomed to endless Muggledom. Perhaps it is just a passing phase. Perhaps the incantation can be ended, the enchantment broken. Raise a glass of cool clear water and toast the possibility!

THE EXTRAVAGANT GESTURE: NATURE, DESIGN, AND THE TRANSFORMATION OF HUMAN INDUSTRY

William McDonough and Michael Braungart

If the landscape reveals one certainty, it is that the extravagant gesture is the very stuff of creation. After the one extravagant gesture of creation in the first place, the universe has continued to deal exclusively in extravagances, flinging intricacies and colossi down aeons of emptiness, heaping profusions on profligacies with ever-fresh vigor. The whole show has been on fire from the word go.

—Annie Dillard

Nature is nothing if not extravagant. Four billion years of natural design, forged in the cradle of evolution, has yielded such a profusion of forms we can barely grasp the vigor and diversity of life on Earth. Responding to unique local conditions, ants have evolved into nearly ten thousand species, several hundred of which can be found in the crown of a single Amazonian tree. Fruit trees produce thousands of blossoms—an astonishing abundance of blossoms—so that another tree might germinate, take root, and grow. Birds, too, seem to have a taste for the extravagant: Who could say the wood duck's plumage is restrained?

For most of our history, the human response to the living earth, to particular places, has expressed the same flowering of diversity. Bearing the unique human ability to imagine and create, we entered the show and developed our own extravagant gestures. We built not just shelter, but beautiful, elegant responses to locale; the breathing, shade-providing Bedouin tent along with the ornate, aspiring temples of cool, coastal Japan. We designed not just wraps against the wind but tailored garments for ritual, celebration, and our own delight. We spoke and moved not just for utilitarian ends but to make drama and poetry, Balinese dance and Shakespearean verse—human creations stoking the fire.

Though human industry in the past 150 years has resorted to brute force rather than elegant design, commerce, too, could become a wellspring of creativity, productivity, and pleasure. Think of the thriving marketplaces that have enlivened the world's great cities, the cherished objects and materials that transform shelter into soulful dwelling. These need not be sacrificed to protect our forests, rivers, soil, and air. Indeed, human industry and habitations can be designed to celebrate interdependence with other living systems, transforming the making and consumption of things into a regenerative force. Design can perform and preserve the extravagant gesture—in the marketplace, in the human community, and in the natural world.

AN AGE OF LIMITS?

For many advocates of sustainable development, the notion that the production and consumption of goods can be a regenerative force is not only alarming, it's downright heretical. Our age is widely perceived as an age of limits. The conventional wisdom holds that the rate of consumption of natural resources by the world's developed nations is damaging the Earth's ecosystems and consigning the Third World to poverty. While some industrialists still use brute force to gain short-term profits, many business leaders have come to realize that a system that takes, makes, and wastes is not sustainable in the long-term.

In response, we all try to limit our impact. We "reduce, reuse, and recycle" at home and in the workplace. Business leaders plan for reductions in resource consumption and energy use. They strive to "produce more with less," "minimize waste," and release fewer toxic chemicals into the air, water, and soil. These industrial reforms, which have come to be known as eco-efficiency, are an admirable attempt to come to terms with the conflict between nature and commerce—they may well help resolve it. But they don't really get to the root of the problem. Working within the same system without examining the manifest flaws in its design, eco-efficient reforms slow industry down without reshaping the way products are made and used. In effect, industry is simply using brute force more efficiently to overcome the rules of the natural world.

Using fewer resources, people may feel a bit "less bad," but no one can quite slip the trap of being merely a "consumer" in a world of poorly designed, toxic products. Every choice seems to contribute to the erosion of human and environmental health: The carpet makes your children sick; the car burns fossil fuels; the TV is loaded with toxic materials. When anything you buy does damage to the world, consumption remains freighted with anxiety and divorced from any notion of sustaining vision that celebrates pleasure, abundance, and delight.

Industry, meanwhile, slogs ahead under regulations that merely dilute pollution rather than examine the cause of the problem; too often these rules are in fact signals of design failure and, ultimately, licenses to harm. And efficiency is proving to fall short of its goals. A new report from the World Resources Institute, for example, announced that pollution and waste in Austria, Germany, Japan, the Netherlands, and the United States have increased by as much as 28 percent in the last twenty-five years, despite increasingly efficient uses of resources. Though Europe in the past ten years has achieved significant reductions in waste, it is merely reaching for sustainability, which is, after all, only a minimum condition for survival—hardly a delicious prospect.

In this atmosphere, the World Business Council for Sustainable Development, a leading advocate of eco-efficiency, found "sustainable consumption" such a burdened, contested term it dropped the word from the title of its four-year study on, well, sustainable consumption. As Ken Alston, a former WBCSD project member said, the group's leadership—executives from major multinational corporations—"struggled to develop eco-efficiency arguments supporting sustainable production and consumption strategies that were robust enough to withstand a critique from environmentalists."

Yet a vision for healthy, sustaining commerce does exist. The idea that the natural world is inevitably destroyed by human industry, or that excessive demand for goods and services causes environmental ills, is a simplification. Nature—highly industrious, astonishingly productive, extravagant even—is not efficient but effective. Design based on nature's effectiveness, what we call eco-effective design, can solve rather than alleviate the problems industry creates, allowing both business and nature to be fecund and productive.

NATURE'S ABUNDANCE

How is it possible for industry and nature to fruitfully coexist? Well, consider the cherry tree. Each spring it produces thousands of blossoms, only a few of which germinate, take root, and grow. Who would see cherry blossoms piling up on the ground and think, "How inefficient and wasteful"? The tree's abundance is useful and safe. After falling to the ground, the blossoms return to the soil and become nutrients for the surrounding environment. Every last particle contributes in some way to the health of a thriving ecosystem. Waste that stays waste does not exist. Instead, waste nourishes; waste equals food.

As a cherry tree grows, it enriches far more than the soil. Through photosynthesis it makes food from the sun, providing nourishment for animals, birds, and microorganisms. It sequesters carbon, produces oxygen, and filters water. The tree's limbs and leaves harbor a great diversity of microbes and insects, all of which play a role within a local system of natural cycles. Even in death the tree provides nourishment as it decomposes and releases minerals that fuel new life. From blossom to sapling to magnificent old age, the cherry tree's growth is regenerative. We could say its life cycle is cradle to cradle—after each useful life it provides nourishment for something new. In a cradle to cradle world—a world of natural cycles powered by the sun—growth is good, waste nutritious, and nature's diverse responses to place are the source of intelligent design.

Industrial life cycles, on the other hand, tend to be cradle to grave. Typically, the production and consumption of goods follows a one-way, linear path from the factory to the household to the landfill or incinerator. Wasted materials and harmful emissions trail products from the cradle of the industrial plant to the grave of the local dump, where products themselves are thrown "away" or burned for energy. Recycling and regulation are often employed to minimize the negative impacts of industry and they do help ease the conflict between nature and commerce. But why not set out, right from the start, to create products and industrial systems that have only positive, regenerative impacts on the world? Why fine-tune a damaging system when we can create a world of commerce that we can celebrate and unabashedly applaud?

Commerce worth applauding applies nature's cycles to the making of things. It generates safe, ecologically intelligent products that, like the cherry tree, provide nourishment for something new after each useful life. From a design perspective, this means creating products that work within cradle to cradle life cycles rather than cradle to grave ones. It means rather than designing products to be used and thrown away, we begin to imitate nature's highly effective systems and design every product as a nutrient.

What is a nutritious product? It's not simply an all-natural product; it's not a recycled product, either. Instead, it's a product designed to provide nutrients to what we have conceived as the Earth's two discrete metabolisms, the biosphere—the cycles of na-ture—and the technosphere—the cycles of industry. Lightweight food packaging, for example, can be designed to be a nutritious part of the biological metabolism; if it is made of organic compounds it can be safely returned to the soil to be consumed by microorgan-isms. Synthetic materials, chemicals, metals, and durable goods are part of the technical metabolism; they can be designed to circu-late within closed-loop industrial cycles, in effect, providing "food" for the technosphere.

Cars, computer cases, washing machines, televisions—in fact, all industrial products—can be designed to retain value as they flow between producer and consumer. Instead of being recycled, or downcycled, into lower-quality materials, products created and used within closed technical cycles—what we call products of ser-vice—can continually circulate as high-quality products. Customers will soon be able to buy the service of such goods, and manufactur-ers will take them back at the customers' request, using their com-plex materials in the product's next high-value iteration.

When products from either the biosphere or the technosphere take a one-way trip to the landfill, a great wealth of nutrients is squandered. Trapped in a plastic-lined dump, organic waste cannot renew the soil and valuable technical materials are lost forever. Worse, the two discrete metabolisms are mixed, contaminating both spheres: Nature, by design, cannot safely absorb the materials of industry and the technosphere has little or no use for organic nutrients. But if the things people make are channeled into one or the other of these metabolisms, then products can be safely manufactured and consumed without straining the environment. They can be considered either biological nutrients or technical nutrients, both of which provide nourishment within their respective spheres of nature and industry.

Our strategy is quite different from the strategy of dematerialization. Proponents of dematerialization aim to reduce the amount of a resource used to create a product. They want to make thinner paper, lighter packaging, a better aluminum can—in this world, less is more. While those innovations may lead to a more efficient use of materials, they do not comprehensively examine the chemistry of materials, the impacts of industrial processes, nor the local circumstances surrounding their use—which may well be quite harmful to both people and nature.

We are proposing something different. We'd like to see a true transformation of commerce, in which design goes beyond using nature efficiently and instead creates value and opportunity with products that nourish rather than deplete the world. This is not to gainsay efficiency. We'd simply like to put efficiency to work in the service of an effective, life-centered vision. As the business genius Peter Drucker has said, being efficient—doing things right—is the crucial role of the manager. It's the leader's job to be effective, to see that the "right things get done." Efficiently managing a toxic system is not the "right thing." Efficient innovations within a life-affirming design protocol, however, suggest a dynamic path to a cradle to cradle world.

FROM MAINTENANCE TO RENEWAL TO INHERENT CREATIVITY

The conceptual, and actual, shift to cradle to cradle products transforms the impact of industry. When all manufactured products and materials are designed as nutrients, the production and consumption of goods enriches the natural world. And when those nutrients flow within coherent cycles, human industry and human desires can become the cherry tree, writ large.

Fanciful? Not at all. Many notable leaders of companies all over the world have begun to move from the maintenance of the old industrial system to a renewal of commerce. They have decided to recognize the far-reaching influence of their creative acts and celebrate their impact on the world rather than disguise it. They have launched the next Industrial Revolution.

In fact, it's already well under way. As early as 1993, the textile industry, led by the Swiss firm Rohner and the textile design company DesignTex, had already developed examples of a textile that is a biological nutrient—a product so safe you could literally eat it. The carpet industry, meanwhile, has adopted the product of service idea and is focusing its business on the concept that carpet can be a technical nutrient retrieved again and again from loyal customers. Both are working to keep their respective materials in coherent, truly cyclical flows.

Companies such as Milliken, Collins & Aikman, and Interface—major commercial carpet companies—are all putting forward their products as materials designed for reclamation. They are telling their customers they want to replace used carpets with new ones and retrieve their technical nutrients. In effect, the companies continue to own the carpet material but lease and maintain it while a customer uses the carpet in their building. Eventually the carpet will wear out like any other, and the manufacturer will reuse its materials in new carpets.

It's important to note, however, that many carpets on the market contain such questionable, potentially toxic materials such as PVC and heavy metals, which cannot be truly "recycled," and are instead shredded and blended into what we call a downcycled material of lower quality—a nylon reinforced PVC mush, for example. Our strategy would imply a redesign of the industry so that carpet materials would maintain their high quality over many useful lives in the technical metabolism.

The chemical company BASF, for example, has recently announced a new fiber called Savant, which is made from an infinitely recyclable nylon 6 fiber. Savant is inherently stain resistant, inherently colorfast—no need for Scotchgard—and designed to be taken back to its constituent resins to become new material for new products. In fact, BASF can retrieve old nylon 6 and transform it into an improved fiber, upcycling, rather than downcycling, an industrial material. The nylon is rematerialized, not dematerialized—a true cradle to cradle product. On the heels of BASF, manufacturers of everything from running shoes to automobiles are designing and implementing new ways to retrieve and circulate valuable materials.

DesignTex, on the other hand, has created an upholstery fabric that flows in the biological metabolism. The company set out to create a product that was beautiful, durable, and ecologically intelligent. After an assiduous design process with the Swiss textile mill Rohner, they decided on a wool-ramie blend that could be removed from the frame of a chair after its useful life and tossed onto the ground to naturally decompose. To ensure that the fabric would safely biodegrade, the design team considered more than eight thousand chemicals used in the textile industry to finish and dye natural fabrics. Most contained some form of mutagen, carcinogen, heavy metal, endocrine disruptor, or bio-accumulative substance, but thirty-eight were found to be suitable for a material destined to be food for the soil.

It was a pleasing outcome: a gorgeous, affordable fabric that would one day be mulch for the local garden club. But the design process also yielded another very positive, if unintended, effect. When regulators tested the effluent from the Swiss mill that produced the DesignTex fabric, they thought their instruments were broken. They tested the influent to check their equipment and found that it was working fine—the water coming out of the factory was as clean as the water going in. The manufacturing process itself was filtering the water.

THE CREATION OF COMMUNITY WEALTH

A textile mill that purifies water begins to suggest the profound impact intelligent design can have on communities. Just as a product designed as a biological nutrient nourishes a community of microorganisms in the soil, a factory and its manufacturing processes can be designed to address a broad range of local concerns, from the desire for a convivial, productive workplace to the health of the environment to the creation of community wealth.

Design creates an environment for a community. A factory designed to nourish a community of workers, for example, can build stronger ties between colleagues by creating pleasant, healthy places for them to work, meet, and enjoy each other's company. That's a laudable intention for workplace design. But it cannot be the only intention. The work community extends beyond the workplace and includes all species that inhabit the locale—not just the human community, but all species. When designers are mindful of all species, their goals change dramatically. Suddenly, the availability of sunlight, shade, and water; the subtleties of climate and terrain; the health of local birds, flowers, and grasses all become fundamental to design. Buildings become responsive to place.

Herman Miller, the furniture manufacturer, took that principle ✓ to heart when it commissioned the design of a 295,000-square-foot factory and office near its headquarters in western Michigan. The company's goals for the new plant were to foster a spirit of collaboration between office and factory workers and create a workplace with a restorative impact on the local environment. Working with a design team that paid close attention to local conditions, Herman Miller built a plant that serves the needs of all its factory workers and administrative employees by celebrating an array of natural and cultural delights.

The low-lying, curved building follows the natural contours of the Michigan grassland. Stormwater spilling off the building moves off the site through an extended series of wetlands that purify the water while providing habitat for hundreds of species of birds, plants, and insects. Plantings of native grasses and trees provided additional habitat for local creatures and further enhance the beauty of the site. Inside the building, offices face the manufacturing plant across a sunlit, urbane promenade, where workers meet and lunch and drink coffee among whimsical sculptures and thriving plants. The entire building—the gyms, the bathrooms, the factory floor—is so pleasantly bright and airy, it is now known as "the greenhouse."

Does this enhance the well-being of workers? Create productivity and wealth? Well, yes. When Herman Miller moved into the building the company was producing $250 million worth of furniture each year. Within a single year it increased production by nearly $50 million with the same number of employees, a gain of 24 percent. At the same time, both office and manufacturing staff reported a significantly higher degree of job satisfaction than they had at their previous workplace.

Herman Miller credits these positive changes to three things: The customized design of the factory, which suited their administrative and manufacturing needs; their innovative management strategy designed to enhance relationships with customers and employees; and the simple fact that the building is such a bright, pleasant place to work.

While it's impossible to measure the influence of delight, its easy to imagine the pleasure of working in a place where you can always see the beauty of the surrounding landscape, where copious fresh air and light actually blur the boundary between indoors and out. Workers in such a place feel as if they have spent the entire day outdoors. They see the comings and goings of birds and the passing of the seasons. They come to know the place where they live during their days at work.

Such pleasures have an enormous impact on the spirit. After Herman Miller moved into the new plant, sixteen young employees left for jobs with higher wages. But they soon returned. When the president of the company asked, "Why are you back?" they said, "We want our jobs back because we had never worked in another factory before. We couldn't work in the dark."

When a company decides to create a workplace where employees can develop an appreciation for local natural beauty, it has given itself the opportunity to rethink everything under the sun; it is making a decision that will ripple through all its endeavors and through the life of the community it inhabits. It is, in effect, making a profound declaration: We are native to this place. For Herman Miller, that meant building a workplace that embodied a new way of thinking about its role in the world. For other companies, for the giants of the Industrial Age, it means staying put, reinventing themselves, and restoring the sites where they have done business for years.

One of those icons of industry, the Ford Motor Company, has launched one of the most sweeping acts of industrial restoration ever. Led by Henry Ford's great-grandson, William Clay Ford Jr., the company has embarked on a twenty-year, $2 billion restoration of its gigantic Rouge River plant in Dearborn, Michigan. Built between 1917 and 1925, the manufacturing complex remains one of the world's largest. At its peak it employed 100,000 workers and churned out millions of cars (boats and airplane engines, too). It was the pride of Ford and the envy of industrialists from Tokyo to Berlin.

Yet, it became a place where workers and management alike worked in the dark: Many of the laborers toiled without seeing the light of day, and management designed and operated products and manufacturing systems with little regard for the natural world. By the beginning of the twenty-first century, the Rouge River plant was a brownfield, a sprawling wasteland of dilapidated buildings, leaky pipes, and old equipment. The land was contaminated, bare of all but the most persistent vegetation, and the river was badly polluted.

Ford Motor could have decided to fence-off the site and build a new factory where land and labor are cheap. Instead, it declared itself native to Dearborn, Michigan. Rather than walk away from a worn-out industrial landscape and a community that had supported it for nearly a century, Ford chose to transform the Rouge River site into a healthy, productive, life-supporting place. Indeed, Ford's leaders are now asking a revolutionary question: When will we be able to let our own children play in the soils and waters of the Rouge?

That critical question leads to a wide spectrum of inquiry. How do we design a manufacturing facility that is a prosperous, supportive work environment? What specific innovations will make the site a place that invites the return of native species? How can the presence of the factory be beneficial to the Rouge River? On the grounds of the site what is the optimum depth of topsoil, number of worms per cubic foot, and insect and bird diversity? What are the optimum aquatic populations of the river?

These may sound like surprising questions for a car company to ask, but Ford is asking them—and answering them, too. Construction began in November 2000 on a new automotive assembly plant that will feature skylights for daylighting the factory floor and a roof covered with growing plants. The 450,000-square-foot "living roof" will provide habitat for birds, insects, and microorganisms. In concert with a series of wetlands and swales, the roof will also control and filter stormwater runoff. With these natural, built-in measures replacing the expensive technical controls called for by new regulations, Ford stands to save between $8 and $35 million on stormwater remediation alone.

Over the course of twenty years—over the course of generations, really—Ford will restore their Rouge River site. In addition to the living roof and the ponds and swales, grasses and other plants will be used to rid the soil of contaminants. Porous paving will filter water through retention beds to further control stormwater runoff. Thousands of trees will be planted to provide habitat for songbirds and, one hopes, beautiful, shady places for the children of Dearborn, Michigan, to play.

These are the kinds of innovations that a manager devoted to efficiency might reject out of hand—too extravagant, too costly. Yet Henry Ford himself, who revolutionized industry with ever-increasing levels of efficiency, would probably have found this an exciting prospect. A plan that invited the return of native species while saving $35 million over conventional engineering, with a delightful landscape thrown in for free, is exactly the kind of cost-effectiveness he would have looked for and insisted on.

As his great-grandson William Clay Ford says, "this is not environmental philanthropy; it is sound business . . . " And he's right of course. Businesses that fail to bring ecological and social concerns to commerce put shareholder value in danger and are not contributing to the larger prosperity. That's why, along with restoring the Rouge River site, Ford is rethinking everything from the materials used in the production of cars to the design of its manufacturing systems.

Indeed, one could reimagine some of the very tenets on which the auto industry has done business for the past hundred years. Bill Ford himself says that the company is no longer simply in the business of building and selling cars and trucks. Instead, they might be in what we call "the personal mobility business." To us this means the auto industry is preparing to design cars as products of service rather than sell them to individual owners. Customers would effectively buy the use of a car for their personal mobility needs for an hour, a day, or a year while the mobility company would provide maintenance and other services. The company would be responsible for their cars forever and would benefit from their valuable materials. A car, in this context, becomes a truly long-term material asset rather than a relatively short-term material liability, and the need to mine the world for raw materials becomes an archaic expression of the industrial age.

That's what separates this from typical leasing: Our consulting firm is working with auto manufacturers who are beginning to imagine building cars that can be completely disassembled and reused. They want to develop everything from new polyesters and paints that retain quality through reclamation to compostable upholstery fabrics that will feed and restore the soil. In the terms of the next industrial revolution, the companies are building a coherent system of closed-loop cycles flowing with technical and biological nutrients.

This is revolutionary. And this lengthy discussion of Ford Motor has simply been to illustrate that the transformation of commerce is already well underway: When an industrial giant with more than $80 billion in purchase orders sends signals such as these to its competitors, its customers, and perhaps most significantly, into the supply chain, one can begin to imagine a wide range of rippling, far-reaching effects; one realizes that the world is changing.

A NEW GLOBAL PERSPECTIVE

The fact that a global company can achieve positive local effects is a very critical issue for us. In our minds, all sustainability is local. On one level, that suggests a rich engagement with one's place, an attitude toward design that draws information and inspiration from the nearby living world. But it can also mean that one develops an appreciation for the distant effects of local actions, and the local effects of distant actions.

When the leader of a large corporation, for example, examines her company's role in the world, she might do so from the narrow perspective of her office. Or she might see that when her decisions initiate labor and create products well beyond her region, they have an impact on a distant place that can only be understood in its local context; what is sustainable in L.A. may not be sustainable in Kerala—or even in New York City. An executive might do nothing with this knowledge, or if she aspires to a sustaining vision, she might begin to take many places into account in her decisions and, in fact, even seek to enrich many places.

Former World Bank economist Herman Daly has approached this idea from a global-economic perspective. He makes a distinction between globalization, a system of uniform rules for the entire world, and internationalization, the increasing importance of relations and trade between nations. While internationalization preserves the identities of nations as it embraces international commerce and communications, in a globalizing economy, says Daly, "what was many becomes one."

From a design perspective, a set of uniform rules for the entire world suggests an erosion of cultural diversity. Applying one-size-fits-all design solutions to architecture, for example, yields bland, uniform buildings isolated from the particularities of place—from local culture and nature to energy and material flows. Such buildings, quite common today in cities and office parks all over the world, reflect little if any of a region's distinctness or style, its unique, often extravagant expressions of humanity.

Consider French cheese. Charles de Gaulle is said to have remarked that it was difficult to rule a country that produced two thousand different kinds of cheese. But should political efficiency overrun diversity? What if the many cheeses of France were to become one? Perhaps that's why the French farm activist José Bové used his tractor to dismantle the McDonald's in his village: for some reason, the thought of such a France seems to have been just too much to bear.

But there's a flip side to the global economy—international trade allows us to experience and celebrate the fullness and diversity of life on Earth. Isn't it to be expected that one might go to a place like New York City to sample the delicacies of Italy and China and Istanbul, all of which are the result of intensely local events? Who would choose to live without Parma's cheeses and hams?

Not the members of Slow Food, an Italian movement working to preserve regional culture with the tools of the global economy. Employing what the movement's founder, Carlo Petrini, calls "virtuous globalization"—a savvy use of global communications to identify international markets for local food producers—Slow Food, writes author Alexander Stille, "has taken up the defense of the purple asparagus of Albegna, the black celery of Trevi, the Vesuvian apricot, the long-tailed sheep of Laticauda . . . and a host of endangered handmade cheeses and salamis known now only to a handful of old farmers."

With the help of Slow Food's commercial ventures—a guide to ✓ local wine and restaurants, a biennial food show—Italian farmers, beekeepers, millers, and vintners are staying in business. A once-struggling miller in the small town of Bra, for example, now has all he can do to keep up with orders for his flour and may soon be grinding grain for the food retailer Williams-Sonoma. That's the beauty of Slow Food: a global network that produces local wealth through a celebration of the pleasures of fine food.

Corporations could also practice virtuous globalization. They might begin by designing products, manufacturing systems, and workplaces that fit the locale. Imagine a global company creating value by applying a high international standard of scientific inquiry—a common tool of corporate research—at the local level, addressing basic needs like nutrition, soil chemistry, or clean water. A prototypical product of consumption such as soap might allow them to do so.

Currently, soap is mass produced and shipped all over the world in a one-size-fits-all solution to a common need. Detergents are designed to lather up, remove dirt, and kill germs anywhere from Brooklyn to Bangkok. Rather than respond to the different washing methods and water chemistries that occur from place to place, manufacturers simply add more chemical force to override local conditions.

That's hardly a benign choice. Though detergent makers proudly announce that their products are "phosphate free," they are not free of other harmful chemicals. The industrial strength required to make a soap work against any contingency makes even a small dose of detergent a potent pollutant. In combination with other effluents in the waste stream, detergents flow into the watershed, diluted but far from safe. The health of rivers and streams, the lives of fish and aquatic plants, the quality of drinking water all take a beating.

There is another way to satisfy the need for clean water and clean clothes. Rather than impose a universal product on markets all over the world, a soap manufacturer might apply sophisticated technology and expertise in chemistry to the development of detergents that are not only safe everywhere, but designed to address the specific needs of ecosystems and deliver nutritious effects to a variety of locales. Soaps for hard water, soaps for soft water, soaps for washing clothes on riverside rocks—even nutritious soaps. Detergents could also be locally produced, providing meaningful local employment, and sold in biodegradable packaging designed to be food for the soil, or in cookie-sized discs, eliminating packaging altogether.

With these innovations, growing organically out of years of research and development, the global company would have developed a product suitable to locale, designed out dangerous chemicals, built an effective delivery system, eliminated waste, protected local waters from pollution, and provided food for local soils. Not bad for laundry detergent.

Many products of consumption are ripe for innovations that will have positive impacts on communities all over the world. A packaging manufacturer could design a biodegradable food container for markets in China, where the disposal of Styrofoam has become a national problem. In India, where waste is often burned for fuel, plastic beverage bottles could be produced with new polymers that would replace dangerous toxins—such as the heavy metal antimony—that are commonly released when incinerated. In fact, polymers produced without antimony have already been designed and offer promising new alternatives in the global marketplace.

IN PRAISE OF DIRTY CLOTHES:
DESIGN AND THE RENEWAL OF EVERYDAY LIFE

If we look at things as simple as soap and water in the context of the daily life of a community, we can begin to see some of the delightful, far-reaching effects of a cradle to cradle world.

Imagine for a moment a community that wants to reinvent itself. After an arduous but exciting round of public meetings, the town's citizens have decided that they want to renew the community's connection to the natural world while restoring the best qualities of a healthy small town. Along with planning to preserve a vibrant commercial district, mixed-use neighborhoods, walkable streets, and lots of parks and playgrounds, the town has also identified the need for a variety of new social venues. Many of these new venues, it is hoped, will bring the generations together in places that provide a pleasing experience of nature during the daily round of errands and chores. In effect, the community wants to take down the fence between commerce, local culture, and the natural world.

One of the new venues is a community service center operated as a viable business by retired people. The center is comprised of a Laundromat, a day care, a health clinic, and a mobility service. It is right on Main Street in the old railway station, which now incorporates new technology to create energy systems powered by the sun, earth, and wind. In fact, high-tech glass, daylighting, photovoltaic panels, and a remote wind turbine in a wind farm off-shore allow the house to operate without a drop of fossil fuel. In a series of indoor botanical gardens and outdoor ponds, wastewater and stormwater treatment is also managed on-site.

Along with the energy and water treatment systems, the old station has received another new addition: a two-story meeting place lit by the sun, a kind of greenhouse commons where elders and infants, parents and teenagers gather at the hub of the neighborhood center. While mothers sit and talk over coffee, enjoying the view of the big, old streetside oaks, a pair of older men relax in the warm sun while waiting for their appointments at the clinic. Others wait to catch the elevator to an underground garage, where a small fleet of community cars is parked.

The community cars are products of service built from re-claimed materials and powered by fuel cells. They are operated by elderly people, who drive around town dropping children at the mid-dle school athletic field, picking up groceries or laundry, and fer-rying people to and from their appointments at the health clinic. The fleet allows most people to keep their cars off the road while giving the community car drivers opportunities to be involved in community life.

Back at the Laundromat, business is brisk and profitable. Most people in the community have chosen not to wash, dry, and iron their own clothes; they have decided its cheaper, takes up less of their time, and is perhaps more ecologically intelligent and socially useful to have their laundry picked up, washed, and delivered each week. So the washing machines are humming—humming with en-ergy provided by the sun and cleaning clothes with detergent that is not simply phosphate-free but completely harmless to the natu-ral world. After each wash, wastewater spills into the indoor botani-cal gardens—creating heat—and then flows outdoors through a con-structed wetland, in each case providing food for local flowers and plants. By design, the community center has become a fecund habi-tat. Like a tree, it makes oxygen, sequesters carbon, fixes nitrogen, purifies water, makes complex sugars and food, and creates a re-storing environment where the generations meet.

From the folding tables in the Laundromat, the views are gor-geous. An elderly gentleman gazes out over the hilly town and after folding a pile of clothes, carries a stack across the commons to the day-care center, where he pauses to watch his grandson play out-doors with the other neighborhood children. He is aware, in the midst of a simple chore, that he lives in a place blessed with an abundance of community wealth and that he contributes daily to its growth.

This is just one example of the many ways in which eco-effective design can transform the experience of everyday life. It suggests how seemingly extravagant gestures—a beautiful social venue for the eldest and the youngest, botanical gardens for purifying water—can add up to a deep sense of community wealth.

Set in the world most of us live in today, such a scenario is shot through with things we might lament, from the impact of fossil-fuel-burning automobiles to the pollution of our rivers and streams to the clear sense that today's industrial strategies will never deliver a high standard of living to all the world's people. We might also ask, given the promise of ecological design, new technologies, and the sensible solutions they enable, why these strategies of change have not been more quickly adopted. Obviously, transformations of the scale we are proposing are not simple. The development and wide adoption of new products, markets, and material-flow systems does not happen overnight, nor without the commitment and energy of leaders from every sector of society. We could imagine this as a long, arduous process. Or we can see that changes of this magnitude, sparked by human need and ingenuity, are a theme of history. As Sheik Yamani, Saudi Arabia's OPEC minister, pointed out during the first so-called energy crisis, the Stone Age didn't end because we ran out of stones.

Human ingenuity responds to the historical moment, and the age of ecologically intelligent design will emerge long before we run out of oil. We believe that it will fully emerge in a cultural shift driven by the engine of commerce, as the values embodied in intelligent design become embedded in the activities of our daily lives. We need only observe the computer revolution to see how quickly technology and economics can make the impossible commonplace. And the shift toward ecological values is already underway. When it becomes widely known, for example, that one of the world's major corporations has chosen to purify stormwater with wetlands and a living roof, and in the process saved $35 million, both business leaders and the culture at large will begin to see the economic viability of ecological design. Once the regulatory infrastructure catches up with designs such as these—designs so inherently productive and safe they don't require regulation—the regulatory agencies will start to use them as benchmarks, presenting them within the culture as strategies that are hugely attractive from the perspective of both the carrot and the stick.

Perhaps just as important, the changes we are proposing allow all of our children a story of hope. Seeking a hopeful future, the tragedies we see will spur many of us to imagine solutions. At first, we may simply try to be free of something we know is harmful, such as chlorine or lead. As we begin to know more about the products we buy, we might make choices about the kind of carpet or mobility system we use based on as much scientific knowledge or personal experience as we can muster. Designers engaged in this transformation would begin analyzing the materials in products and replacing harmful chemicals with more benign ingredients. Other products would be designed only from materials fully defined as safe biological or technical nutrients. And at the highest level, designers would begin to develop systems to assemble products in ways that allowed for their coherent, cradle to cradle flow within the technical and biological metabolisms.

These changes are within our grasp. Indeed, they have already begun. Innovations in architecture and community design are being employed all over the world. The revolution in product design is well under way. And as we begin to realize the fruits of our efforts, today's laments will become celebrations of a world in which people and nature thrive together—abundantly, delightfully, extravagantly . . . hopefully.

IN SEARCH OF JUSTICE

Rep. Nydia M. Velázquez

Jacob Riis's groundbreaking 1890 book *How the Other Half Lives* chronicled the lives of thousands of families living in squalor in New York City. A horrified public cried out for reform, and the public-housing movement was born. Over the next fifty years, a bold progressive movement brought extraordinary change in America—an income tax began to bridge the gap between rich and poor, a social-safety net provided basic protections for workers, and we began to understand that a just society was a social and moral imperative. What happened?

Today, although we have a much more sophisticated understanding of what social justice means, low-income families nevertheless find themselves in more or less the same relative circumstances they were in over a century ago—struggling to overcome desperate poverty, living in communities that are tragically unhealthy as the result of pollution and neglect, and laboring against a culture that considers material possessions the absolute measure of social value.

The simple fact is that our current unsustainable "more-is-better" culture undermines any hope of achieving justice—at home or abroad. We often hear about how the United States consumes a vastly disproportionate amount of resources relative to the rest of the world. Americans are building bigger houses, driving bigger cars, consuming more and more of everything than just about anyone else anywhere.

This is certainly true, and the long-term environmental effects of this overconsumption may well prove disastrous. But we also forget that the gap between the rich and the poor in this country is just as severe as, if not worse than, it is elsewhere. Amazingly, the richest 196 people in America have more wealth than the poorest 56 *million* Americans.

And one thing is for sure—Americans certainly are not doing all this overconsuming in congressional districts like the one I represent. The residents of Greenpoint/Williamsburg, which makes up the heart of New York's Twelfth Congressional District, are among the poorest in the country. Forty-five percent of the households earn less than $12 thousand annually. In some schools, nearly every child qualifies for the federal school lunch program.

My district is by no means unique. Across America, there are pockets of dire poverty that are a national disgrace. Perhaps this poverty is not on the scale of some developing-world countries, but it is crushing poverty nonetheless. The U.S. Department of Health and Human Services reports that nearly three million American children suffer from moderate to severe hunger. More than nine million children report having difficulty obtaining enough food, suffer from reduced quality of diets, express anxiety about their food supply, and are increasingly resorting to emergency food sources and other coping behaviors. And nearly seven million Americans are classified by the U.S. Department of Labor as being the "working poor," by spending more than twenty-seven weeks in the labor force, but earning below the official poverty level.

In my district, crime is high, test scores are low, schools are crumbling, and the "American dream"—however you choose to define it—is very, very difficult to attain. For many, hope is represented by a trip to the bodega to buy a lottery ticket. Side by side with the "more is better" dominant culture is an unnoticed "anything is too much" underclass that scrapes for crumbs in the shadows. How can a nation with our riches allow such misery to exist?

AN ENVIRONMENTAL CATASTROPHE

One of the first things you are likely to notice when you visit neighborhoods in my district, and others like it, are the trucks. Not trucks heading out, packed with manufactured goods produced in clean, modern, high-tech facilities full of good-paying jobs. No, the trucks are delivering garbage from the rest of the city to waste-transfer facilities located cheek by jowl with schools, apartment buildings, and small businesses struggling to keep afloat. It's illegal to put a cigarette billboard near a child's school. It's perfectly legal to have a garbage dump next to her house.

Amazingly, the waste transfer stations aren't the worst of it. The tour of environmental shame continues on to highly toxic empty lots—so-called "brownfields." Usually, brownfields are the remnants of polluting industries sited in poor communities whose owners have fled, leaving virtually undevelopable property in their wake. Cleanups do occasionally occur, at enormous public expense, while polluters walk away, often to set up shop in another unwitting community. Political opponents may cry "class warfare," but I've personally never seen a waste transfer station on the upper East Side of Manhattan, or in the Hamptons. In some communities, residents complain about the presence of federal buildings, which generate no real estate tax. We'd take a federal building or two in the twelfth district in a New York minute.

The environmental scorecard in my district is appalling. There are over 80 waste-transfer stations in New York City, and 30 of them are in my district. Greenpoint/Williamsburg has 137 sites that use hazardous substances, called right-to-know sites, 15 toxic-release inventory sites, 24 waste-transfer stations, and one low-level radioactive waste site, all within a five and a half mile radius. Red Hook has 6 waste-transfer stations, 7 sites that may warrant Superfund designation, and 3 hazardous waste facilities, all within a one-mile radius. Greenpoint/Williamsburg sits on a seventeen-million-gallon oil spill (The Exxon Valdez, by contrast, spilled ten million gallons of oil. Of course, they at least tried to clean that one up.) The Newtown Creek Sewage Treatment Plant treats the highest volume of hazardous waste and is the biggest producer of hazardous pollutants in the city. The neighborhood is also home to 8 coal-burning industrial furnaces and the highly toxic Brooklyn Navy Yard (a federal facility we'd gladly give up).

Hardly a day goes by when we don't hear about how our nation's cities have experienced a renaissance. Although crime rates have dropped in many large cities, people of color are being victimized by toxic polluters and brownfield perpetrators who have managed to escape the wrath of a well-meaning environmental movement that doesn't seem to realize that the "environment" does not begin at the suburb's edge. Believe me, Latinos and African Americans know what the environment is. It's air they can see and water they can't drink. Polling consistently shows that African Americans and Hispanics put a higher priority on protecting the environment than nonminority voters. And no wonder. Study after study has shown that racial minorities disproportionately bear the brunt of this pollution.

- A 1992 EPA study showed that minority populations are disproportionately exposed to air pollutants and hazardous-waste sites.

- A 1990 University of Michigan study showed that minority residents are four times more likely to live within one mile of a commercial waste facility than whites, and that race was a better predictor of proximity to such sites than income.

- A 1994 study by the Center for Policy Alternatives concluded that three out of every five Latinos and African Americans live in a community with one or more toxic-waste site.

The effect of this pollution is a silent national tragedy on a par with Jacob Riis's alarming discoveries. In the predominantly African-American area between New Orleans and Baton Rouge, dubbed "Cancer Alley," where over 130 petrochemical plants, medical waste incinerators, and solid-waste landfills emit highly toxic levels of pollution, people suffer from much higher rates of miscarriage, cancer, tumors, and other chemical-related illnesses. A study by the Louisiana Advisory Committee to the U.S. Commission on Civil Rights found "Cancer Alley" to have disproportionate levels of all cancers. The Latino south side of Tucson is exposed to twenty times the acceptable levels of trichloroethylene, and rates of cancer, birth defects, and genetic mutations in that neighborhood far outpace national averages.

I don't have to walk very far in my district to see a community in the throes of a health crisis. New York City has the second worst air quality in the nation, and the Greenpoint/Williamsburg area has the worst air quality in New York. And it is no surprise that the people who live in this neighborhood are getting sick and dying at an alarming rate.

A 1993 New York Department of Health study showed that in one census tract, childhood cancers are twenty-two times higher than the national expected average. Twenty-three census tracts showed disproportionately higher rates of stomach and lung cancers. Four showed statistically elevated incidences of leukemia. Hundreds of cases of lead poisoning in children have been reported in the past decade. Childhood asthma rates in Brooklyn have tripled in the 1990s. At 13 per 1000, the infant mortality rate rivals that of Estonia, Bulgaria, and the Czech Republic.

I have a constituent whose cousin died of an asthma attack on an Ozone Alert day—at age twenty-two. Her two-year-old daughter now lives with her grandparents. Another constituent's grandchildren suffer regular asthma attacks when they come to visit, only to see the attacks subside when they leave.

It's outrageous that basic environmental-justice protections languish in Congress. There are those in Washington who claim that they want to make sure that "no child is left behind," yet the hollowness of the claim leaves a deafening echo.

It goes without saying that the communities most adversely affected by the actions of these polluting industries don't have the resources to fight back, and it's little wonder that the many waste-treatment facilities in Williamsburg process garbage from outside the neighborhood.

The residents of Greenpoint/Williamsburg and other poor communities, fighting disease and living in poverty, are no match for wealthy companies who have an unlimited ability to litigate. Unless we bring the full force of the federal government to bear on these polluters, they will continue to poison our air and water and slowly and silently subject our children to a toxic environment with impunity. Where is Jacob Riis when we need him?

We know that protecting the environment is a long-term necessity and that global-climate change puts future generations at risk, but we need to understand that the environment is a matter of life and death right now in our inner-city communities. As our children are slowly poisoned, their ability to learn and to work and to become active, productive participants in this society is being taken away. We are quite simply writing off millions and millions of American children because they happen to be poor.

A DESPERATE SOCIAL CHALLENGE

If the environmental impacts of being poor in today's America are bad, the social effects may be even worse, if that's possible. Those Americans who drew the short straw and live in poverty are systematically shut out of the blessings of American society, Horatio Alger success stories notwithstanding. Talk is cheap in Washington, and talk about "values" is cheaper still. If we really valued work, then the janitors and garbage collectors and sweatshop workers and the rest of the hardworking poor would be able to put food on the table. If we really valued children, we'd make sure that the poorest of our children weren't taught in hallways and broom closets or in shifts and we'd guarantee that they all had textbooks and qualified, well-paid teachers.

A poor child in today's America labors against a dizzying array of social challenges. He lives in a culture that says that what you have is more important than who you are. The message from advertisers and marketers who target kids is quite clear—you gotta have it. And it doesn't really matter what the "it" is. From clothes to computers to cell phones to houses to cars, no American child is immune from the underlying suggestion that owning these *things* defines success. While the message of excess materialism is toxic for all our children, it is especially cruel for the one out of six American children living in poverty. I often wonder how sporting goods executives sleep at night after marketing basketball shoes to low-income children that cost a minimum-wage–earning parent nearly a week's salary to pay for.

Our role models are athletes and movie stars and billionaires. How can we expect our poorest children, malnourished and undereducated, to feel good about themselves when we present these false notions of success? Our real role models should be neighborhood volunteers, nurses, teachers, parents, and small business owners, and all those caring, hardworking people who create jobs, watch over us, and give a community a sense of cohesion, connectedness, and purpose. One of the few good things to be taken from September 11 was the emergence of blue-collar heroes. Let's hope it lasts, but the smart money is still on Michael Jordan and Britney Spears.

It might be possible to overcome even these formidable obstacles if we gave our children some basic building blocks for success, but we don't. Poor public school districts are reduced to opening up our schools to advertisers and retailers who plaster the walls with ads and line the hallways with vending machines for junk food and soft drinks. In exchange for televisions and VCRs, millions of public school children are forced to watch commercial television programming from the for-profit Channel One television network. One study concluded that taxpayers in the United States pay $1.8 billion per year for the class time lost to Channel One.

Parents are torn—they think that the choice is to either allow corporations to commercialize our public schools or to do without necessities. They've been conned. Right now, we're seeing an appalling lack of political will to provide necessary resources to our schoolchildren. When the call came to fight terrorism, we found the money. Democrat or Republican, we did our best to toss partisanship and politics aside to do what needed to be done. But when some of us in Congress try to build support to fight ignorance and illiteracy, to educate the next generation of entrepreneurs and innovators, and to provide the tools to free an entire class of Americans from poverty, we're told, "There's no money, let 'em watch TV."

BLUEPRINT FOR A JUST SOCIETY

People fighting for justice in this country and around the world have every reason to be discouraged, but in fact there are signs of hope that we should be aware of. Anyone familiar with my work in Congress knows that I don't give up easily. If anything, I've just gotten started. Even now, in the face of extraordinary challenges, I am excited about the emerging opportunities in my community and others like it. There are three primary areas that I have dedicated myself to that could help make huge strides toward true social jus-tice—environmental protection for low-income communities, eco-nomic opportunity for small business, and a true commitment to a basic safety net of core social protections that includes providing a quality education for all children.

Environmental Justice

To its credit, the Clinton Administration acknowledged what those of us familiar with such communities knew for years—that people of color have been unfairly treated with regard to the devel-opment, implementation, and enforcement of environmental laws, regulations, and policies.

Under former administrator Carol Browner, the Environmen-tal Protection Agency established the Office of Environmental Jus-tice, whose mission is to address the problem of environmental discrimination and set up programs designed to assist what it ac-knowledges are "disproportionate risks faced by ... low-income and minority populations." Under the Bush Administration, the program saw its already meager budget reduced.

In 1994, President Clinton signed Executive Order 12898, which applies Title VI of the Civil Rights Act to entities receiving federal financial assistance and bars them from using methods or practices that discriminate on the basis of race, color, or national origin. It was a good start to begin to undo generations of wrongs that have been visited upon people of color.

One of the first fights I took up in Congress was to ensure that minority communities receive equitable treatment under environ-mental law. I did this by introducing legislation to apply this nation's civil rights laws to our environmental regulations. If enacted, my legislation would take the former president's Executive Order an important step further by applying Title VI to actions taken by any company whose pollution disproportionately impacts a particular racial or ethnic group. Title VI is most commonly used in discrimina-tion cases involving public housing and education, but there is no reason that it couldn't be extended to environmental concerns as well.

After all, can anyone say that the health crisis that currently affects people of color in this country is any less important than acts of housing or education discrimination currently protected under Title VI? My proposed legislation targets both potential violators and chronic polluters. Companies would have to show that proposed projects would not pollute in a manner that would adversely affect a particular racial group—the burden of proof currently used under Title VI. Existing polluters would be forced to prove that their pollution does not target minority communities.

My sense is that the effect of environmental justice laws could also ripple through the manufacturing sector. Producers might just find themselves dematerializing their production process if they no longer were able to dump waste in the nearest poor neighborhood. They might just start making cleaner products if they were unable to release toxic chemicals with impunity. The nascent green revolution could get a much-needed shot in the arm from an unlikely ally—the justice movement.

While a law to ensure environmental justice would make a tremendous difference for the lives of millions of poor people in this country, it's honestly not enough to just keep polluters out of our communities. To have true environmental justice, we also have to provide safe, open space, parks and community gardens that make neighborhoods worth living in. In my district, there is a desperate struggle underway over the fate of the abandoned Brooklyn waterfront. This was once a highly active shipping and manufacturing center, and if you look past today's rotted piers and disrepair, you can't help but see the promise it holds. With its easy access to transit and a breathtaking view of Manhattan across the East River, the waterfront is perhaps New York City's most potentially valuable real estate. Developed properly, with cooperation from the community and visionary investors, we could create a truly inspired vision of a sustainable community that incorporates open space, creates opportunities for nonpolluting businesses and the jobs that go with them, and that could become a model for urban infill development across the country. This is an opportunity to build an urban development project applying principles of sustainability and combining ecological and economic objectives.

The Small Business Miracle

When I came to Congress in 1993, it became obvious to me that we would have to find new solutions to spur economic development in low-income communities. The old model of large-scale manufacturing is a thing of the past in my district and others like it, but this is not necessarily a bad thing. As we have seen, these manufacturing operations were invariably highly polluting, and were often run by distant owners with little stake in the lives of their employees and the communities in which they lived. Large-scale layoffs had a devastating effect on workers, as owners simply left town with the profits and shipped their operations overseas. It seemed to me that if we could encourage small entrepreneurs to open businesses, we would have a real chance of providing badly needed jobs and services that could help people emerge from poverty. I immediately sought a seat on the House Small Business Committee, and I now serve as the Democratic leader. Our committee oversees what I feel is the most dynamic sector of our economy, and the one that holds the most promise for promoting justice.

Small businesses are the engine of this economy. They account for half of the gross domestic product and create more than three-quarters of all new jobs. They often lead the country out of difficult economic times. Many of the world's largest businesses—including Disney, Hewlett-Packard and Microsoft—all got their start as small businesses in the midst of economic recessions. Small businesses spend their profits locally, contribute to vibrant, diverse local economies, and are quite simply the social and economic glue that makes a town or neighborhood worth living in.

There are, needless to say, any number of obstacles to starting such an enterprise in a low-income community. We're beginning to establish programs that overcome these obstacles, and the results are extremely encouraging. We've been able to open Small Business Development Centers, which provide vital technical assistance and counseling, and we've expanded programs that assist women-owned businesses, the fastest growing sector of the business community (women are starting businesses at about twice the rate of men, and there are now more than nine million women-owned businesses employing more than twenty-seven million workers and generating annual sales of more than $3.5 trillion). We've opened counseling centers for entrepreneurs in low-income communities, we're setting up business incubators for converting derelict warehouses into clean, modern commercial space, and we're beginning to overcome perhaps the most serious obstacle to success—lack of access to start-up capital—through the New Markets Venture Capital Program.

This is just a start, but it's an extremely good start. We're already learning a great deal from these programs, and we're now able to apply the lessons learned in ways that are benefiting communities around the country. The investments in microloan programs, small business incubators, and technical assistance centers are minimal compared to the benefits. Communities with thriving small businesses have lower crime rates, lower unemployment, and a greatly reduced need for public assistance. We're creating solid economic opportunities for people who are rooted in their communities and who will be the next generation of social, economic, and political leaders.

Building a Just Future

If we want to create a truly just society, here are the building blocks: (1) guarantee a quality education for all children, regardless of their neighborhood, class, or race; (2) provide safe, affordable child care so that the parents of at-risk youth can work; (3) ensure basic health care; (4) guarantee a living wage for all workers. From there, the work we are already doing on environmental justice and community development can take off.

Would these suggestions cost a lot of money? It depends on how you count. We have shown a willingness, if not an eagerness, to spend money when we believe it to be in our national interest. What can be more in our national interest than eliminating poverty and hunger and ensuring that every child born in this country has a real opportunity to contribute?

More than that, though, these solutions are good for all of us.

Transforming our society into a place that values people over material goods gives our lives meaning. Taking the necessary steps to make sure that everyone has the right to a clean environment, a decent job, and a safe, thriving community lifts us all up. Across the country and especially in New York, the generosity, bonding, and community building that took place in the aftermath of September 11 helped remind us that taking care of those in need makes us better as a people. If we build on that model, we have a real chance to create a society we can all be proud of.

CLEANING THE CLOSET:
TOWARD A NEW FASHION ETHIC

Juliet B. Schor

I love clothes—shopping for them, buying them, wearing them. I like good-quality fabrics, such as wool or linen. I cultivate long-term relationships with favorite items, such as sweaters and scarves. I delight in a beautifully tailored suit, everything perfectly in place. And I love to find a bargain.

I confess these sartorial passions with some trepidation. Love of clothes is hardly a well-regarded trait by my friends in the environmental, simplicity, feminist, labor, and social justice movements. And for some good reasons. Much of what we now wear comes from foreign sweatshops. Textile production, with its toxic dyes, often poisons the environment. Fashion is a sexist business, which objectifies and degrades women. Young people adopt a must-have imperative for the latest trendy label. Adults have problems too: The typical compulsive shopper, deep in credit-card debt, has been supporting a shopping habit focused mainly on clothes, shoes, and accessories. Even a cursory look at the making, marketing, wearing, and discarding of clothes reveals that the entire business has become deeply problematic. But my friends have other objections that I find less compelling. Many believe that clothes are trivial—not worth spending time or effort on. Some feel they are irreversibly tainted by the excessive importance society has placed on them, or the power of the greedy behemoths that dominate the industry.

One school of thought—call it "minimalist"—takes a purely utilitarian stance. Clothes should be functional and comfortable, but beyond that, attention to them is misplaced. The minimalist credo goes like this: Buy as few clothes as possible, or better yet, avoid new altogether, because there are so many used garments around. Make sure your garments don't call too much attention to themselves. Shun labels and "designers." Purchase only products whose labor conditions and environmental effects can be verified.

This ethic has gained its share of adherents in recent years. A growing number of young people critique their generation's slavish devotion to Abercrombie, NorthFace, and Calvin, preferring the thrift-shop aesthetic. Simplifiers advocate secondhand stores, clothes swapping parties, and yard sales. The market for organic clothing, despite its generally inferior design and high prices, is expanding. *No Logo* has developed its own cachet.

Clothing minimalism is certainly a morally satisfying position. But most people do not and will not find minimalism appealing, and not because they are shallow or fashion addicted. Rather, minimalism fails because it does not recognize the centrality of clothing to human culture, relationships, aesthetic desires, and identity. Ultimately, minimalism lacks a positive vision of the role of clothing and appearance in human societies.

But what could that positive vision be? First, it will affirm the cultural importance of clothing, rather than trivialize it. It will embrace the consumer who buys conscientiously and sustainably, but who also has a prized and beautiful wardrobe hanging in her closet. It will recognize that apparel production, which after all has historically been *the* vanguard industry of economic development, should provide secure employment for millions of women and men in poor countries, a creative outlet for designers and consumers, and a technological staging ground for cutting-edge environmental practices. A "clean-clothes" movement has begun in Europe. Can we transform it into a "clean and beautiful clothes" movement here in the United States? If so, it holds the potential to become a model for a wider revolution in consumer practices. For if we can work it out with a commodity as socially and economically complex as clothing, we can do it with anything.

CLOTHES BY THE POUND

For an introduction to the insanity of the industry, a good place to start is a used-clothing outlet. I chose the Garment District, a hip, department-style warehouse in East Cambridge, Massachusetts. Inside, it's chockful of every retro and contempo style one could imagine. Outside, huge eighteen-wheeler trucks deposit giant, tightly wrapped bales of clothing, gleaned from charities, merchandisers, and consumers. These clothes sell for a dollar a pound, and seventy-five cents on Friday. That's a price not too much above beans or rice.

Over the fourteen years that the Garment District has been in business, the wholesale price of used clothing has dropped precipitously, by 80 percent in the last five years alone, according to one source. Renee Weippert, director of retail services at Goodwill International, reports that prices in the salvage market dropped to two to three cents per pound by late 1999 and are now in the seven to eight cent range. In this supply-and-demand oriented "aftermarket," the price decline has been caused by an enormous increase in the quantity of discarded clothing. Throughout the 1990s, donations to Goodwill increased by 10 percent or more each year.

And what of the clothing that is not resold to consumers? The Garment District sells its surpluses to "shoddy mills," which grind up the clothes for car-seat stuffing and other "post-consumer" uses. Or they send it into the global used-clothing market, where it is sold by brokers or given away by charitable foundations. Ironically, the influx of cheap and free clothing in Africa, under the guise of "humanitarian aid," has undermined local producers and created more poverty. And that's not the only irony—the excess clothing that ends up in Africa, the Caribbean, or Asia, probably also started out there.

THE POINT OF PRODUCTION:
SWEATED LABOR AND THE POISONED LANDSCAPE

Textiles have become the vanguard industry in the emergence of a new global sweatshop, where women—who comprise 70 percent of the labor force—work for starvation wages, making the T-shirts, jeans, dresses, caps, and athletic shoes eagerly purchased by U.S. consumers. The brutal exploitation of labor and natural resources is at the heart of why clothes have become so cheap.

Consider the case of Bangladesh—which by late 2001 was the fourth largest apparel exporter to the United States. The country, with a per capita income under $1,500 per year, a 71 percent female illiteracy rate, and 56 percent of its children under age five suffering from malnourishment, is one of the world's poorest. While proponents of corporate globalization claim the process is lifting people out of poverty, a recent study by the National Labor Committee reveals otherwise. Wages among Bangladesh's 1.6 million apparel workers range from eight cents per hour for helpers to a high of eighteen cents for sewers. Workers are forced to work long hours and are often cheated of their overtime. When demand is high, they work twenty-hour shifts and are allowed only a few hours of sleep under their sewing machines in the dead of night. The workers, most of whom are between sixteen and twenty-five years old, report constant headaches, vomiting, and other illnesses. Even the "highest" wage rates meet less than half the basic survival requirements, with the result being that malnutrition, sickness, and premature aging are common. Ironically, apparel workers cannot afford to buy clothing for themselves—a group of Bangladeshi women factory workers who recently toured the United States report getting only one new garment every two years. The university caps they sew sell for more than seventeen dollars here; their share is a mere 1.6 cents per cap.

These conditions are not atypical. Disney exploits its Haitian workers who make Mickey Mouse shirts for twenty-eight cents an hour. Wal-Mart, which controls 15 percent of the U.S. market and is the world's largest clothing retailer, has Chinese factories that pay as little as thirteen cents per hour, with the norm below twenty-five cents. High-priced designers also exploit cheap labor—Ralph Lauren and Ellen Tracy pay fourteen to twenty cents, Liz Claiborne twenty-eight cents. Nike, despite years of pressure by activists, continues to exploit its Asian workforce. At the Wellco and Yue Uen factories where its shoes are made, the company was paying only sixteen to nineteen cents per hour, requiring up to eighty-four hours per week including forced overtime, and employing child labor. A recent estimate for a Nike jacket found that the workers received an astounding one half of a percent of its sale price; a study of European jeans found a mere 1 percent went to workers. By contrast, "brand profit" accounts for about 25 percent of the price.

But low wages are only part of the horror of the global sweatshop. Many factories and worker dormitories lack fire exits and are overcrowded and unhealthy. Workers sewing Tommy, Gap, and Ralph Lauren clothes have been found locked inside the factories. They are routinely harassed—sexually and physically—by their supervisors. The Bangladeshi workers report that beatings are common and that they are forbidden to speak inside the factory. Permission to go to the bathroom is severely curtailed, many are forced to work while ill, and companies typically fire those who become pregnant. Unions are bitterly resisted, with terminations, physical harm, and intimidation by employers. The retailers who contract with local factories have tried to build a wall between themselves and their subcontractors, but this is little more than a callous ruse.

Manufacturers are also exploiting the natural environment. While clothing is not typically thought of as a "dirty" product, like an SUV, plastics, or meat, a closer look reveals that this clean image is undeserved. From raw material production through dyeing and finishing, to transport and disposal, the apparel, footwear, and accessories industries are responsible for significant environmental degradation. Consider cotton, which makes up about half of global textile production. Cotton cultivation is fertilizer-, herbicide-, and pesticide-intensive, endangering both the natural environment and agricultural workers. The crop comprises only 3 percent of global acreage, but accounts for 25 percent of world insecticide use. In some cases, the crop is sprayed up to ten times per season with dangerous chemicals, including, among others, Lorsban, Bladex, Kelthane, Dibrom, Methaphos, and Parathion. The toxicity of these chemicals ranges from moderate to high and has been shown to cause a variety of human health problems, such as brain and fetal damage, cancer, kidney and liver damage, as well as harm to birds, fish, bees, and other animals. Not surprisingly, farm workers suffer from more chemical-related illnesses than any other occupational group. Chemical run-off into the nation's drinking water has also been extensive—Aldicarb, an acutely toxic pesticide, has been found in the drinking water of sixteen states. Conventional cultivation also depletes the soil and requires large quantities of irrigation water.

Additional hazards arise from chemical-based dyeing and finishing of cloth. The most common chemical dye, used in textiles and leathers, is the so-called azo-dye, which is now believed to be carcinogenic and has been banned in Germany. Formaldehyde, pentachloraphenal, and heavy metals remain in use despite their toxicity. A little-known aspect of these toxins is their human health impact on both workers and consumers. One German study found that 30 percent of children in that country suffer from textile-related allergies, most of which are triggered by dyes. An estimated 70 percent of textile effluents and 20 percent of dyestuffs are still dumped into water supplies by factories. In South India, where the (highly toxic) tanning industry grew rapidly in the 1990s, local water supplies have been devastatingly polluted by large quantities of poisonous wastes. The various stages of textile production (from spinning, weaving, and knitting to dyeing and finishing) also require enormous energy and water use. For example, one hundred litres of water are needed to process one kilogram (2.2 pounds) of textiles.

Environmental effects can also be more indirect. The consumer rage for cheap cashmere has led to unsustainable expansion of herds in Mongolia and, subsequently, to overgrazing, desertification, and ecological collapse. The growth of new fabrics, such as the wood-based tencel, is contributing to deforestation in Southeast Asia.

Environmental impact does not end at the point of production. The globalization of the industry has led to increased pollution through long-distance transport. And eventually, the products enter the waste stream. Clothing, footwear, and accessories are a staple of municipal landfills.

SUPERFLUITY, NOVELTY, AND EXCLUSIVITY: HALLMARKS OF THE CLOTHING INDUSTRY

At the core of the disposal problem lie two developments: Clothes are cheap and Americans are buying them in record numbers. Since 1991, the price of apparel and footwear has fallen, especially women's clothing, with the drop especially pronounced after 1999. (This was most likely due to declining wages in Asia, caused by the Asian financial crisis.) It is no surprise that as clothes got artificially cheaper, Americans began accumulating more of them. Indeed, when prices are low, the pressure on manufacturers and retailers to sell more becomes intense. In 2000 alone the United States imported 12.65 billion pieces of apparel, narrowly defined (i.e., not including hats, scarves, etc.). It produced another 5.3 billion domestically. That's roughly 47.7 pieces per person per year. (Women and girls' rates are higher; men and boys' lower.) From Bangladesh alone we imported 1.168 *billion* square meters of cloth. That's a lot of caps.

Paradoxically, the system of low prices and high volume is anchored at the top by outrageously priced merchandise. At the high end, thousand-dollar handbags, dresses running to the many thousands, even undergarments costing a hundred dollars are the rule. A look at the nation's distribution of wealth provides one clue to why high-priced clothing is flying off the shelves: The top 10 percent of the population now own a record 71 percent of the nation's total net worth, and 78 percent of all financial wealth. (The top *one percent* alone own 38 and 47 percent of net worth and financial wealth.) The existence of such an upscale apparel market is a troubling symptom of a world in which some people have far too much money and far too little moral or social accountability in terms of what they do with it.

But the high-priced venues serve another purpose as well. Designer merchandise becomes available at discount stores at a fraction of its top retail price. This affordable exclusivity is part of what keeps middle-class consumers enmeshed in the system. Clothes cascade through a chain of retail outlets, prices falling at each stage. The system has led many consumers to purchase almost mindlessly when confronted with irresistible "bargain basement" prices of highly regarded designers and to spend much more on clothes than they intend or even realize. Eventually even the desirable designer merchandise ends up being sold for rock-bottom prices—on the web one can find surplus clothing sites selling clothes at a fraction of their retail prices. I found $5,000 designer dresses going for $1,000, women's coats that retail at $129 available for $22 each; men's down jackets for $12. I found Hilfiger,

DKNY, Victoria's Secret. Brand-new "high-quality mixed clothing" can be had for twenty-two cents a pound.

The core features of contemporary fashion—fast-moving style, novelty, and exclusivity—also contribute to spending. A seasonal fashion cycle based on climactic needs has been replaced by a shorter timeline, in which the "new" may only last for two months, or even weeks, as in the extreme cases of athletic shoes. The exclusivity that is relentlessly pushed by marketers also contributes to high levels of spending—the product is valued *because* it is expensive. As it becomes more affordable, its value declines. Similarly, when the consumer aspires to be a fashion pioneer, she seeks rarity. The impacts of these core features of the fashion industry are profound. Many middle- and lower-middle class youth are working long hours to buy clothes. For poor youth, with limited access to money and jobs, the designer imperative has been linked to dropping out of school (because of an inadequate wardrobe), stealing, dealing, even violence. Failing to keep up with the dizzying pace of fashion innovation undermines self-esteem and social status.

But it is not only fashion-orientation that accounts for the enormous volume of clothing that is sold in this country. Shopping for clothes, footwear, and apparel have become habits, even addictions, especially for women. Just something to do because we do it. People shop on their lunch hours, on the weekend, through catalogs, or in the mall. They spend vacations at outlet malls. Americans typically have something in mind they want to buy, and for women that something is often clothing. In my interviews of professional women who subsequently downshifted, a common refrain was the enormous superfluity of their closets. It's clear we need to get our relationship to clothes under control.

WHY CLOTHES *DO* MATTER

To create sustainable, humane, and satisfying apparel, footwear, and accessories industries, we need to understand the functions of these products. Their utilitarian features are obvious. We need garments to cover our bodies, hiking shoes to climb a mountain, a watch to tell time. But this is just the beginning of what clothing really *does*. Throughout history, clothing has been at the center of how human beings interact. Not always with humane purposes, of course. Clothing and footwear have long identified rank and social position. Before the nineteenth-century, European governments passed sumptuary laws that regulated dress, particularly to control those of low status. Intense conflict was waged over whether one could wear a wig, choose a certain color, or sport a particular fashion style. Not surprisingly, clothing was equally central to struggles against those very inequities. Working people often asserted their social rights by choosing dress that elites deemed them unworthy of wearing. Clothing has been key to both the repression of social groups and their struggles for human dignity and justice. (Closer to home, consider how incomplete any account of the political challenges of the 1960s would be without attention to blue jeans, tie-dyed shirts, and long hair.)

Clothing has also been at the heart of gender conflicts. As early-twentieth-century American women attempted to break free of patriarchal strictures, they rejected corsets and confining dresses. In the twenties, they defied convention with cigarettes and short skirts. In the 1970s, the women's movement rejected the fashion system and created its own sartorial sensibility. Clothing has historically also been an important site of intergenerational bonding and learning between women, along with hair care and other beauty rituals. I am quite sure that I got my love of clothes from my mother. Being a good shopper, especially in the complex world of women's clothing, requires finely honed skills, which are passed on from generation to generation.

What we wear is important to the way we experience our sexuality. Our age. Or ethnicity. It allows us to show respect for others (by dressing specially for a social occasion) or to signal community (through shared garments or styles). Finally, clothing can be part of the aesthetic of everyday life. There is genuine pleasure to be gained from a well-made, well-fitting garment. Or from a piece of clothing that embodies beautiful design, craftspersonship, or artistry. Throughout history, human beings have exercised their creativity through clothing, footwear, and accessories.

In sum, dressing and adorning are a vital part of the human experience. This is why any attempt to push them into a minimalist, utilitarian box will fail. Clothes embody far more than our physical bodies; they are also a measure of our basic values and culture. So, while we may not all take great pleasure in what we wear, we should all recognize that clothes *do* matter. They are about as far from trivial as any consumer good can be. Which means that a new fashion ethic will be about affirming social and human values, the commitments of daily life, and our hopes and aspirations for a different kind of world.

PRINCIPLES FOR A NEW KIND OF CLOTHING CONSUMER

1. Quality Over Quantity: Moving from Cheap and Plentiful to Rarer and More Valuable

In the past, clothing cost more, and its use was far more ecologically responsible. Only the rich bought more clothes than they wore, in contrast to current habits. Expensive clothing was worn sparingly when newly acquired. In some places, as a garment wore out, it cascaded through a social hierarchy of uses, from esteemed social occasions to the everyday public, and eventually to the most mundane private and domestic uses. Clothing was cleaned far less frequently than today, thereby extending its useful life. Women had the skills to make clothes, and even as ready-made garments became more common, to restructure and upgrade them. Style could be attained through "refashioning" garments, rather than discarding them and buying new ones. Such refashioning could also involve new ownership. Historian Neil McKendrick has identified the origins of the eighteenth-century consumer revolution in Britain with the trickle-down from elites to servants, as maids took their mistresses' cast-off dresses and turned them into newly stylish outfits. Thus, basic principles of ecology and frugality were maintained—take only what you need, use it until it is no longer useable, repair rather than replace, refashion to provide variety.

The history of clothing practices provides guidance for fashioning a new aesthetic whose central principles are to emphasize quality over quantity, longevity over novelty, and versatility over specialization. For example, if we reject the need to keep up with fashion and can be satisfied with a smaller wardrobe, we can spend more per garment, as consumers do in Western Europe. The impact on the earth is less, and it contributes to longevity, because better clothes last longer by not skimping on tailoring or quality and quantity of yardage. Consumers are better off because high-quality clothing is more comfortable and looks better.

Ultimately we could begin to think of clothing purchases as long-term commitments, in which we take responsibility for seeing each garment through its natural life. That doesn't mean we couldn't ever divest ourselves, but that if we grew tired of a useful garment we'd find it a new home with a loving owner, kind of like with pets. Of course, to facilitate such a change, consumers would need to reject the reigning imperative of variety in clothes, especially as it pertains to the workplace and for social occasions. Just because you wore that dress to last year's holiday party doesn't mean you can't show up in it again.

With such an aesthetic, consumers would demand a shift toward more timeless design, away from fast-moving trends. Clothes could become more versatile in terms of what they can be used for, their ability to fit differently shaped bodies and to be altered. Consider the Indian sari, a simple, rectangular piece of cloth that is fitted around the body. It accommodates weight gain and loss, pregnancy, growth, and shrinkage. Couldn't designers come up with analogous concepts appropriate to Western tastes? Pants with waistbands that are flattering but also can be adjusted through double-button systems or through tailoring. Basic pieces that can be complemented by layering and accessories. Expensive, classic clothes already have some of these qualities—extra fabric for letting out and the capability to remain flattering after they have been altered.

Striving for longevity through versatility facilitates what we might call an ecological or true materialism. The cultural critic Raymond Williams has noted that we are not truly materialist because we fail to invest deep or sacred meanings in material goods. Instead, our materialism connotes an unbounded desire to acquire, followed by a throwaway mentality. True materialism could become part of a new ecological consciousness. Paying more per piece could also support a new structure of labor costs. Workers would work less, produce fewer but higher-quality items, and be paid more per hour. Such a change would help make ecologically clean technologies economically feasible.

Finally, paying more for clothes does not mean adopting the premise of social exclusivity. In luxury retailing, much of the appeal of the product is its prohibitive price or the fact that only elites have the social conditioning necessary to pull off wearing it. An alternative aesthetic would value democracy and egalitarianism through the fashioning of garments that are high-quality but affordable.

2. Small and Beautiful: Creative Clothing for Local Customers

The aesthetic aspect of clothing is and will continue to be important. But the values represented by the fashion industry are unacceptable. Despite decades of feminist criticism, the industry continues to objectify women—and increasingly men—through demeaning, violent, and gratuitously sexualized images and practices. In the late nineties we got "heroin chic," glamorizing drug abuse and poverty. Now it's teen and "tween" styles, with bare midriffs, tightly fitting T-shirts, and sexually explicit sayings emblazoned on the garments. Furthermore, the industry is comprised of megacorporations employing a small number of mostly male designers. They, in turn, produce a monolithic fashion landscape—massive numbers of copycat garments. Suddenly all that's available are square-toed shoes, or short-handled handbags, or hip-huggers.

An alternative vision starts from the recognition that many young people, especially young women, yearn to be fashion designers, producing garments that are artistic, interesting, funky, visionary, and useable. And consumers are increasingly desirous of that type of individualized clothing. The industry could return to its roots in small-scale enterprises, run by the designers themselves. The British cultural analyst Angela McRobbie has envisioned such a shift, calling for small apparel firms located in neighborhoods, operating almost like corner stores. They would cater to a local clientele whose tastes and needs they come to know. These face-to-face relationships between female designers and the immigrant women who labor in domestic-apparel production also have the potential to reduce the exploitation that currently characterizes the industry. Instead of driving to a mall with its cookie-cutter stores, one might walk to a converted factory housing three or four designers with workshops-cum-showrooms. The consumer could also become active in the creative process, helping to fashion an interesting or unique look for him or herself. If she didn't see what she wanted, it could be made to order, so that fit, color, and style were just right. Such a system would yield substantial savings in the areas of transport, branding, advertising, and marketing as well as a dramatic reduction in overproduction. Those savings could be used to pay decent wages, install environmentally sustainable production technologies, fund better quality materials, and support designers.

Such a vision could be realized through a combination of activist pressure, consumer mobilization, and government policies. The federal government could offer special subsidies for training and education for designers and enterprise loans to small business owners. Local governments could support apparel manufacturers through tax incentives and marketing initiatives.

3. Clean Clothes: Guaranteeing Social Justice and Environmental Responsibility

Relocalization is an important part of a movement toward a just and sustainable apparel industry. But it must go hand in hand with improvements in wages and working conditions in factories and small production units abroad. Such reform is essential to relocalizing on a global scale, because it will be the foundation for creating purchasing power in India, China, Bangladesh, and other southern countries. For now, the north must continue importing in order to provide employment for impoverished foreign workers. But as wages rise abroad, these workers can produce for their own domestic markets.

One of the most important social movements of the past decade has been the coalition of labor, student, and religious activists opposing the exploitation of garment workers around the globe. The Gap, Nike, KMart, and others have been exposed and embarrassed by their labor practices. Students have demanded that their college's insignia clothing not be produced by sweated labor, and more than ninety institutions have complied. Most American consumers now believe that the workers who make their clothing should be paid decently, and surveys indicate they are willing to pay somewhat more to achieve that goal.

To date, however, the industry response has been inadequate. While some progress has been made, far more energy has gone into winning the PR battle than has been devoted to substantive reform. Companies remain opposed to free association in unions, which is the only true long-term solution to abuse. Nevertheless, the principle of what Europeans call "clean clothes" is making headway. In Europe, major clothing retailers have committed themselves to codes of conduct that ensure reasonable working conditions, free association, and other labor rights. For example, the British chain Marks and Spencer has joined the Ethical Trading Initiative, which is a government-sponsored initiative bringing together nongovernmental organizations (NGOs), unions, and businesses. Next, another British chain, works with Oxfam on ethical trading.

Indeed, the successes of the European clean-clothes movement are worth looking at, particularly for extending beyond labor rights into environmental impacts. In 1996, the Dutch company C&A instituted rigorous controls over its suppliers—monitoring more than one thousand production units annually—to guarantee labor conditions and environmental impacts. It uses the Eco-Tex label for environmental certification, and many of its own brands sport it. Marks and Spencer has begun an organic cotton design project with the Royal College of Art.

The German company Otto Versand, the largest mail-order business in the world, has perhaps gone farthest in terms of environmental sustainability. It has reduced paper use in its catalogs and packaging; its mail-order facility uses wind and solar power; and it is moving to incorporate sustainability throughout its product lines. Otto subsidizes the production of organic cotton in Turkey and India, and last year offered 250,000 organic cotton products. The company has reduced the use of harmful chemicals in textiles and has certified that 65 percent of its clothing passes a strict "skin-test" for dangerous substances. In the late 1990s, Otto worked with Century Textiles (India's largest textile exporter), to phase out azo-dyes. The company has also introduced its Future Collection, which is oriented to production ecology through conservation of energy and water resources. To encourage consumers to adopt a long-term perspective, they offer a three-year replacement guarantee for all their clothes.

To be sure, the shift to just and ecologically sustainable clothing is not simple. The price of organic clothing is currently high, putting it out of reach for many consumers. But activist pressure can help solve this problem, as the European successes are showing. And the U.S. market is already increasing. Nike and The Gap have begun to use some organic cotton. If one or two major U.S. companies commit to a substantial program of organic cotton use, demand will grow and prices will fall. And even a high-priced company such as Patagonia has made some accommodations for affordability—all its clothes carry a no-questions-asked indefinite replacement guarantee and the company operates a number of discount outlets.

The successes of the European campaigns suggest that comparable progress is possible on this side of the Atlantic as well. For example, Eileen Fisher, a high-end women's retailer, has signed on to SA 8000, an international social and environmental standard. U.S. manufacturers and retailers are sensitive to the need to maintain their public image. If we can educate consumers and mobilize activists, we can "clean" the American closet. Doing so would be a substantial step toward a sustainable, but also fashionable, planet.

CHANGING THE NATURE OF COMMERCE

Jeffrey Hollender

In the heart of the city of Burlington, Vermont, lies a paradise. Here, where the Winooski River completes its task of draining distant mountain rains into Lake Champlain, the ages have carved a tangled floodplain of marsh, forest, river, and field known as the Intervale.

Even today, the Intervale remains a world apart, a small but enduring universe of cattail and heron, oak and blackbird. In a sense, it is a place where the past faces the future, where what has been confronts what is to come and attempts to forge a coexistence in which both might gain some measure of prosperity from the other.

It is fitting, then, that in this landscape where natural history flows toward tomorrow an experiment will soon begin. It's an experiment called the EcoPark, and in its unfolding is a vision of the way the world soon must work.

In the words of its designers, the EcoPark represents an agriculture-based industrial ecology. The businesses that will occupy the four-acre facility will form a largely self-sustaining commercial community in which waste, as in nature, is nonexistent.

At the philosophical heart of the EcoPark is the idea of zero impact. The wastes of one tenant will become the feed stocks of the next. The inefficiencies of modern industry, seen so clearly in the vast quantities of effluent now discharged into our air, soil, and water, will not be present. Instead, a Living Machine, a labyrinth of enormous interconnected water tanks each containing a marine ecosystem, will transform these wastes into useful things.

At the EcoPark, grains grown in adjacent fields will be used by a resident brewer. The brewer's byproducts will become raw materials for a resident baker. The baker's wastes will become organic fuel for the Living Machine, which will cleanse the park's wastewater and produce fish for the city. Overflow organic wastes will be composted with waste paper and used to fertilize the facility's input crops.

Drawing energy from waste heat produced by a nearby wood-powered generating plant and incorporating a variety of other similar interdependencies within and without its boundaries, the Eco-Park intends to be a model of sustainability, one in which virtually all negative environmental impacts created by its businesses are transformed into useful things.

Complementing this approach to its environmental capital, the EcoPark's designers are wisely planning to create a supportive environment for its employees as well. Fair wages, progressive policies, and outstanding working conditions will be a foundation upon which an uplifting work experience is built.

In many ways, the EcoPark will be everything that our current global system of business is not. It is a logical response to present commercial circumstances that, for the most part, cannot be maintained, and at least a partial resolution to the conflict that underlies this unsustainability.

Today's businesses are largely driven by two factors: the idea of limitless growth and a preoccupation with short-term results. While neither of these philosophical foundations is right or wrong in and of itself, both are at odds with the fundamental organizational principles of the natural world in which businesses must exist. Unlike most companies, nature is predicated on finite limits and clearly manages its affairs with the longest possible term in mind.

Too many companies, on the other hand, consume natural resources as if there were a limitless supply. With constant pressure to increase quarterly earnings, not much thought is given to the judicious use of manufacturing inputs and other consumables, or to the implementation of more ecologically minded processes, because it's felt such changes could have a negative impact on profits. At the same time, waste products, many quite hazardous, are being copiously produced and largely deposited wherever it's cheapest. In most cases, this ultimately means our air, soil, and waterways. When it comes to the products and services themselves, many are frivolous or even harmful in one way or another, built with inferior materials, and/or deliberately designed to need relatively quick replacement. In the broadest terms, business is basically converting the natural wealth of the earth into eventual garbage far faster

than the earth is able to replace those resources or absorb the results of their consumption. As economist Herman Daly asserts, under our present economic system, the earth is a business in liqui-
dation.

From a human standpoint, our system of business is unsustainable as well. Workers are treated as just another input to be used
until its exhaustion necessitates replacement. From Third World
sweatshops to urban office towers, people are too often forced to
endure the lowest financial, emotional, and physical conditions
that a company can get away with. Instead of caring for employees,
companies hope to be able to largely ignore them. An individual's
aspirations, the dreams that make them who they are, are usually
deemed unrelated to the corporate cause. The result is a poisonous
downward spiral of diminishing expectations and negative behavior
on both sides of the equation, a vicious circle that finally leaves all
parties convinced the only option is to accept this fate. Such patterns are simply not sustainable.

The good news is that because businesses themselves are the
driving force behind all of these trends, they also have the power to
reverse them, and in fact such changes have begun as more and
more companies come to appreciate the virtues of responsible
business practices and the need for new ways to seek growth in a
world rapidly running out of room.

The theory that businesses can and should be a force that aggressively promotes greater environmental and social good has its
roots in the social transformations of the 1960s. Though that era
was marked by a certain innocence, it was also a time of meaningful
idealism. The explosion of social and environmental awareness that
defined a generation led to a great willingness to explore possible
new paradigms. As the young people seeking these personal alternatives matured, they sought better paths professionally as well
and put their values and spirit to work creating a different kind of
commerce.

In the last thirty years, this lasting belief in the need to create
a new business model has given rise to an emerging breed of commercial entity called the socially responsible or sustainable business. Embodied by such pioneering companies as The Body Shop,

Patagonia, and Ben & Jerry's, this new brand of corporation is based on something Ben Cohen of Ben and Jerry's has labeled "caring capitalism," an initially eye-raising blend of old-fashioned money-making and committed human and environmental altruism, and a resulting dual bottom line in which profits are measured in ways both old and new.

While private enterprise has always supported select charitable causes, and while some companies have worked to create workplaces that empowered individuals and minimized environmental impacts, this new business model represented something significantly different. Here were companies building such ideas into their business plans and designing their operations around the belief that there exists an explicit commercial duty to place human and ecological welfare on an equal footing with financial success.

This new wisdom has been a long time coming. Although sustainable businesses are now well established, and though many of the concepts they embrace have moved into the mainstream business world, in their purest and most necessary form they remain something of a curiosity.

At the same time, however, commerce has assumed an ever-increasing presence on the global stage. At its own insistence, the business world has asked for, and in some instances simply usurped, societal decision-making rights on a par, in many cases, with government. The result has been an unprecedented power to shape our lives, influence our cultures, and steer history itself.

For these and other reasons, the time has come to move beyond the notion of individual sustainable businesses and toward the idea of sustainable business, a new totality in which all companies adopt a doctrine of social and environmental responsibility.

There exists for this transformation a blueprint of sorts, a map made by the companies that have beaten the first paths to sustainability through the commercial wilderness. But these revolutionary efforts are a beginning, not an end. There's a pressing need to look past them to a new and wholly reimagined universal system of commerce predicated not upon the sacrifice of nature nor the subjugation of human beings, but upon the enhancement of both.

This is the vision of the EcoPark, and to get to it we must do as its creators have done and reawaken ourselves to fresh possibilities.

This process begins with a question: What is sustainability? At its core, sustainability is the theoretical ability to exist in perpetuity. A sustainable endeavor operates for the long haul. Its activities do not result in the eventual demise of the individual systems upon which the whole depends for survival. A sustainable enterprise, then, becomes one in which vigilant respect is paid to its environmental and human resources, and all activities are structured in such a way as to enrich these elements.

This idea, of course, is easier to put on paper than into practice. However, the difficulties involved in the translation to reality do not in any way preclude the possibility of a fully sustainable system of international commerce. For its own sake (if nothing else) the business community must see sustainability as more than one possibility among many. It must see it as a probability that must be achieved if business is to succeed in any lasting way at all.

The quest for a universally sustainable business system must simultaneously take two parallel paths: one that leads to restored global environmental health and one that leads to a materially secure and spiritually enriched population. We cannot have one without the other.

Let us begin with the more urgent of these tasks: an elimination of all negative impacts on the environment. Simply put, we must stop damaging the biosphere that supports life because that damage is approaching the point of no return.

The evidence is everywhere: Bloated with carbon dioxide, our atmosphere has produced the fourteen warmest years on record in just the last twenty-two. Nearly 20 percent of the world's forest cover has been lost since 1700. Estimated extinction rates are at least 100 times higher than in premodern times. Synthetic organic chemical production has increased 3000 percent since 1940, and the bodies of every single man, woman, and child now carry dozens of these toxins. Such are the results of the industrial revolution and the consumer habits it has fostered.

To reverse these trends, we must seek a zero emissions standard for all commercial enterprises. This is a tall order, and business alone does not currently possess the motivation to develop the technologies we'll need. For that reason, the government must assist by funding environmental technology research programs to speed the process. Before we do anything else, we must imagine a world where all air, water, and soil are free of contaminants and come together to forge the technologies that will build it.

Such an effort need not start from scratch. Indeed, many organizations have been working on such ideas for years. The United Nations–based Zero Emissions Research and Initiatives (ZERI) offers a leading example. According to its mission statement, ZERI is to undertake scientific research . . . with the objective of achieving technological breakthroughs that will facilitate manufacturing without any form of waste. ZERI founder Gunter Pauli says that ZERI is "driven by the desire to keep the body of this Earth healthy," and indeed anything less than a zero-emissions industrial system will not contribute to this goal. Fortunately, ZERI feasibility studies have concluded that such a system is quite possible, and since 1996 some fifty ZERI-sponsored projects have been providing concrete proof.

The 3M Company's Pollution Prevention Pays program (3P) is one pioneering attempt to see waste and pollution as economic inefficiencies that decrease profits and damage the environment. Since 3M officially acknowledged these essential points in 1975, the program's results have been impressive. The company has prevented the release of some 1.5 billion total pounds of pollutants into the environment. Waste releases into local waterways have dropped 75 percent from 6.4 million pounds in 1990 to 1.6 million pounds in 1998. A model water-treatment facility in the company's Singapore electronics plant goes even further by separating the various elements of the facility's waste stream for proper diversion. Ninety-nine percent of all metals, for example, are removed and many are recycled, 50 percent of exiting water is pumped back inside for reuse. Though not a closed-loop system, the 3P program is nonetheless impressive and has saved the company $810 million.

That's no small point. Contrary to popular dogma, preventing emissions is a profitable pursuit. Companies make money when they develop and license technologies that end waste. And companies that use these technologies enhance their profit margins as well because such technologies increase manufacturing efficiency. Pollution, after all, is the result of a raw material being incompletely used, and raw materials cost money.

Where waste and waste products cannot be eliminated, options to simple disposal or environmental release must be found. We must build an alternative waste stream in which waste products are converted to another use. In nature, nothing is wasted. Every output is an input for something else. Our system of commerce must adopt an identical strategy and process its effluents via a closed-loop system in which wastes are commodities.

In some cases, this is simply not possible and here material substitutes must be found. Dioxins, for example, an inadvertent class of by-products of many chemical processes involving chlorine, have no known use yet rank among the most toxic and persistent substances ever produced. In situations like this, we must not accept outputs like dioxins as a necessary evil of an industrial society, chemicals whose ubiquitous presence in our bodies represents a sacrifice we make in the name of progress. Instead, we must find alternatives to the materials and processes that create them or do without such things.

Yet even in the case of dioxins, solutions abound. The two key sources of this toxin are chlorine paper bleaching and waste incineration. In a closed-loop society, waste incineration will not be necessary. And paper can be safely bleached with hydrogen peroxide. This readily available alternative, already in use at several far-sighted paper mills, essentially breaks down into oxygen and water. Problem solved.

Pollution, however, isn't the only destructive output of commerce. There are also its products and services to consider, and these must be subjected to a life-cycle analysis that examines the environmental effects of each component from creation to use and disposal. Nontoxic substitutions for toxic materials and processes must be found. Products must have recycling built in, and de-manufacturing facilities must be created where all consumer products are reclaimed.

Such green-product design is already a way of life for many companies. At AMP, Inc., a manufacturer of electronics, all plastic components are stamped with identification symbols to facilitate their recycling. The company's Design for Environment program trains its engineers to aggressively seek environmentally safe materials and techniques in all facets of product development.

Some companies have even gone so far as to turn a product into a service. Interface, Inc., an Atlanta-based manufacturer of commercial carpeting, operates an innovative program in which they rent rather than sell their product. The company installs and maintains carpets for customers, replacing worn sections as needed and reclaiming all materials for reuse. Safety Kleen, a vendor of cleaning products, pursues a similar strategy, leasing solvents instead of selling them in order to reclaim unused quantities.

At Seventh Generation, the company I operate, we're beginning to build a life-cycle analysis into our product development. We ask questions about the materials we use: Where do they come from? What is the environmental cost of their extraction and processing? What are the working conditions like for the people that make them? When it comes to the products themselves, what are the consequences of their use and disposal?

In the case of our paper and cleaning products, the historical answers weren't good. Household tissue papers are largely derived from forests, subjected to chlorine bleaching, and sold with an avoidable excess of unrecycled packaging. In the case of cleaning products, ingredients are predominantly sourced from petroleum. Extraction and refinement of this resource damages the environment, and the resulting substances themselves are often extremely toxic. Packaging faired no better: Overly large containers were made from virgin plastic that was sometimes nonrecyclable.

Our reengineered alternatives addressed these problems. We developed paper products made from 100 percent recycled paper with an ever-increasing amount of post-consumer content. Our paper is either unbleached or bleached with peroxide to prevent the release of toxins, and some rolls are compressed to allow for decreased packaging, which itself is made from recycled materials where currently possible. The results are significantly more sustainable paper goods.

For our cleaners, we worked with chemists to develop formulas based on renewable, vegetable-based resources instead of petrochemicals. The ingredients we define as acceptable must pose no toxic threat during use and must biodegrade harmlessly. At no point during the life of our products can the environment or user be exposed to potential harm. (Even given all this, there's always room for improvement. Our cleaners should eventually be made from certified organic ingredients. And our packaging needs to be 100 percent post-consumer recycled, among other things.)

In addition to healthier products, the sustainable business will also eliminate the idea of planned obsolescence, in which products are designed to have a limited lifetime so as to foster replacement purchases. Instead, products must be built for the longest life possible and designed to be cheaper to repair than replace. Easily exchanged components, more durable materials, and hardier engineering must end the current era in which it makes more economic sense to buy a new VCR for $100 than fix a broken one for $150.

There is also the crucial issue of inputs, or those materials and resources a business consumes. Inputs, from energy to office paper and production materials, must be shifted from unsustainable resources to those that are renewable and recycled.

When it comes to energy inputs, businesses are vastly more wasteful than need be. Energy-efficient lighting and motors and other technologies exist that can dramatically reduce a company's power usage and expenses without sacrifice. Even the simplest retrofit can pay quick dividends.

In the late 1980s, the Pennsylvania Power and Lighting Company upgraded a single twelve thousand square foot workspace with energy-efficient lighting and a new design that maximized illumination. The cost was over $8,000, but energy bills fell by 69 percent and productivity rose over 13 percent, which garnered an additional $42,000 in yearly savings. When energy savings, productivity gains, and lower maintenance costs were combined, the payback time on the initial investment was just sixty-nine days, operating costs fell 73 percent per year, and the initial investment produced a startling 540 percent return. The impact of every business in the country adopting these available technologies would be equally staggering.

Similarly, heating and cooling upgrades, intelligent architecture, energy-efficient office equipment, and other strategies could, if used universally, cut annual commercial energy consumption and expenses in half and prevent a pollution output equal to six million cars. Big numbers. Small changes. Let's make them.

Of course, we'll still need power. Just not the kind we use today. Instead, we must shift our entire energy system to sustainably renewable sources. This is not nearly as difficult as it sounds. If the federal government funded energy research in the same way it funded the Marshall Plan or the space program, the job could be accomplished quickly.

In addition to funding development initiatives, governments must also alter their economic and environmental policies to support sustainable technologies. Indeed, it's fair to say that businesses will not be able to achieve meaningful sustainability without significant changes in these areas.

Too often there are fundamental policy conflicts within governments that send mixed signals to the marketplace. In the United States, for example, federal purchasing policies mandate the use of recycled paper even as the Department of Agriculture encourages the clear-cutting of virgin forests through massive subsidies to companies extracting timber from public lands. Such conflicts must be eliminated so that government policies and programs, from taxation to subsidization, coherently foster sustainability.

Similarly, both business and governments must redesign their accounting systems so that they measure the human and environmental costs of doing business. These very real expenses are completely ignored by the generally accepted accounting principles, used to calculate everything from the Gross National Product to corporate P&L statements. As a result, there exists no bottom-line incentive to alter unsustainable practices or products. What must replace this system is something economist Herman Daly calls "full-cost accounting," a new method that adds previously ignored expenses to a revised bottom line in order to create powerful arguments for their elimination. Such expenses should include everything from environmental damage to related human illnesses. If, for example, the prices of cars included the costs of cleaning up the air pollution they emit and healing those made ill by manufacturing discharges, huge changes would quickly come.

Policy and accounting changes would also create strong new incentives for a necessary conversion to a new "carbohydrate economy" that replaces our use of hydrocarbon feedstocks for everything from fuel to plastics with plant-based materials. We can already create cleaner fuels from plant-based ethanol and researchers are working on reliable plastics made from corn and hemp. Given our civilization's increasing technological prowess, there's no reason to believe we can't identify a host of other economically viable, biobased alternatives to unsustainable fossil fuels and petrochemicals.

Whenever the business sector fails to pursue ideas like these, regulations should be put into place requiring or stimulating their adoption. And on that score, both business and government should cease and desist with the tired jobs vs. the environment argument. There is absolutely no evidence that environmental standards cost the economy and every proof that they enhance it.

According to a recent study by the Worldwatch Institute, for example, building an environmentally sustainable economy has created fourteen million new jobs around the world and holds the promise of creating millions more. These new economic opportunities are a direct result of recycling and remanufacturing, renewable energy development, and energy and materials efficiency. The report found that job losses due to environmental regulations amount to less than 0.1 percent of all layoffs in the United States, and, in fact, jobs are more likely to be jeopardized in areas where environmental standards are low and development of clean technologies is lacking. Similarly, the Institute for Southern Studies found that states with the best environmental records also offer better job opportunities and economic climates. The study concluded that economic and environmental protection are strongly linked and that states (and, I would add, businesses) do themselves no economic favors by sacrificing the environment for short-term gain.

The other, equally important side of sustainable business practices concerns the human dimension. Just as commerce must respect the environment, so must it respect its employees. A company is composed of its people. Together, they create its profit, and businesses should acknowledge all individual and collective contributions to its cause.

The final mark of any advanced organization, be it a civilization or a company, is that it displays genuine kindness toward the people within. But even if we set aside such high concepts for a moment, it's clear that there's a component of self-interest at play here. A wise company realizes that happy people are motivated people; exactly the kind that create innovation and success.

To achieve that end, and because it's the right thing to do, a sustainable company must manage its human resources in a wholly compassionate way and engage in practices that promote both personal fulfillment and professional development among its employees. Employees must be seen as equal partners, valued for the contributions they make and respected as human beings. This respect must take the form of livable wages and equitable labor practices. But it should also extend beyond the basics of salary and a safe work environment.

At Seventh Generation, we want the work experience of our people to be the best one they've ever had. That's a big job, and it's not always easy, yet for the sake of both our employees' well-being and our company's success, it's not something we want to overlook. This concern takes many forms. On a physical level, our facility provides everyone with generous amounts of natural light and fresh air. Weekly massages are offered onsite, and all employees get a health club membership, to name just a few examples. Our efforts to ensure emotional satisfaction run a similar gamut. We set aside everything from funds to pay for additional education to dedicated time at staff meetings to gauge everyone's well-being and find out how to further enhance it. Practices like these have resulted in a workforce that, for the most part, is as dedicated as it is happy. They come from our recognition that our company is only as sustainable as the health and spirit of its people. Simply put, we ignore these elements at our personal and professional peril. A sustainable business also actively engages the communities in which it conducts business. It's not enough for a company to provide jobs and pay taxes. That's a start, but certainly not an end. Again, this is true if only from a perspective of self-interest: If the community a company does business in is unhealthy, the company could become similarly unsound. No business, after all, exists in a vacuum, and the success or failure of each is influenced by the condition of the world around it. To ensure mutual prosperity, businesses should seek a positive proactive relationship with their neighbors.

If, for example, local schools are failing, nearby businesses would likely soon face an inadequately educated labor pool. Encountering just such a situation in Los Angeles, Universal Studios created a program called Education is Universal. This partnership with local school systems provides money, scholarships, paid volunteers, and mentors to schools, at Universal's expense, in order to promote learning and career development among L.A. students. The community gets an improved education system. Its young people get a shot at a better future. Universal gets better employees. Everybody wins when a company collaborates with its community in the drive to make that community a better place to live.

The path to sustainability must extend to a company's business partners as well. Companies should do everything they can to influence suppliers, distributors, dealers, franchisees, licensees, agents, and customers. They must see themselves as positive catalysts spreading their standards throughout the system. Taking this a necessary step further, companies should monitor business partners for their compliance with these standards and be willing to cease doing business with any that continues to willfully engage in unacceptable practices.

At many companies, this influence is already a way of life. IBM, for example, works with vendors to develop stringent environmental standards. Decisions about purchases and ongoing business relationships take into account a supplier's environmental performance. Outdoor clothing manufacturer Patagonia took the concept to a new level in the mid-1990s with the company's decision to switch to organic cotton for its natural fiber clothing. Though not without costs to the company and its customers, the switch caused many suppliers to rethink their own product lines.

A similar trickle-down theory of sustainability should be practiced internally as well. A company's management team must seize every opportunity, whether by example or training, to instill in all employees the values they wish their company to embody. In this way, a necessarily pervasive culture of sustainability can be created.

The importance of such a culture cannot be understated. If a company's efforts to achieve genuine sustainability across all facets of its operation are to be successful, it must realize that true sustainability is the sum of countless individual employee actions and decisions, some small, some not so. Together, these behaviors combine to create that which the company is. Or is not.

These factors make aggressive internal education a key part of the sustainable paradigm. At my company, we've created a six-day training program for new employees, two entire days of which I conduct myself. While this is a huge investment, the returns are incalculable. In effect, we're compressing into a single week the six to twelve months it would otherwise take new hires to naturally acquire sustainable instincts. This steep learning curve allows new employees to independently embody our company's values almost immediately, a circumstance that keeps our company's quest for ultimate sustainability from faltering even during periods of rapid growth.

Education efforts must also extend to a company's customers. Marketing materials should educate as well as sell. The added sustainable value of a company's goods and services must be made clear because most consumers remain largely uninformed about the environmental and social factors involved and make purchasing decisions with little or no information beyond price. The sustainable business needs to lead the way in educating consumers about the importance of these other attributes with marketing that seeks to describe both the larger problems and solutions to them.

When a company undertakes marketing-based education, consumer behaviors are transformed in several crucial ways. First, consumers begin to see the broader issues behind the sustainable business movement and start to appreciate the need for systemic change. With this appreciation comes support for the company's efforts to spur that change and a likely willingness to pay a little more for its products or services in the process. In the end, such efforts quite often create the two things sustainable businesses need most: Additional voices applying new pressures on the unsustainable system itself and customers who generally display more gratitude and subsequent loyalty than most.

One need look no further than the organic-foods industry to see these principles in action. Not long ago, organic foods were an obscure niche product largely ignored by most. Over the last decade, however, industry education efforts have both convinced consumers of their importance and successfully persuaded them that they're worth paying more for. The result? Industry sales are growing at nearly 23 percent per year, and expected to double to $20 billion annually in the next four years. With mainstream food companies now jumping on this accelerated bandwagon, our entire food system is being slowly reengineered. And public education is largely responsible.

These are just some of the opportunities available to businesses seeking sustainability. In adopting them, however, companies must also be careful not to change certain behaviors. Indeed, the core value of financial success must be preserved if the sustainable business model is to become demonstrably viable.

One of the biggest mistakes socially and environmentally committed entrepreneurs make is focusing on the more admirable aspects of their mission at the expense of the economic considerations that make a business tick. Many excellent ideas have withered on the vine because more attention was paid to the kind of paper in the copy machine than the amount of cash in the bank. As I've seen many times, if one moves too quickly on sustainable fronts, it becomes exponentially easier to destroy the business itself.

Employees from the CEO on down must always pay vigilant attention to the business itself and be aware that this focus will frequently result in what I've come to call the art of the intentional compromise; those delicate decisions made amidst the tensions that lie between a business's desired reforms and its need for profitability.

My company's paper products are a perfect example. Because post-consumer recycled fibers cost more than either virgin paper or pre-consumer fibers, and because 99 percent of the public doesn't really appreciate the environmental difference, we're constantly managing a tension between recycled content and price. Sometimes we have to let our post-consumer content fall rather than charge ourselves out of existence via prices that are beyond understanding or acceptability. At the final analysis, it's far better to make a slightly less environmentally benign product for a time than it is to find yourself out of business and unable to make any difference at all.

Ultimately, businesses that are strongly working to revamp their practices so as to support human and environmental well-being will be rewarded with a greater market share created by growing consumer demand for sustainable goods and services. This will, in turn, allow them to push the sustainable envelope even further. In the short term, however, the trick is achieving a balance between moving too fast and not moving fast enough. In running a sustainable business, this has been the source of my greatest struggle: How do you juggle your interests without completely sacrificing any one of them? How do you know when to make compromises? When to trade some short-term sustainability for long-term financial security or vice versa? Every company must carefully consider how to integrate its sustainability with financial success, and one will sometimes cost the other. But which one? And when?

The answer, to borrow from Gertrude Stein, is that there is no answer. It's a game of trade-offs in which one hopes to take two steps forward while taking no more than one back somewhere else. I see it very much like writing a piece of music. There are no rules that say you have to start here with violins or end there with piano or that the rhythm has to be just so. It's an artistic process in a sense, one filled with surprises, choices, and unexpected shifts in focus.

My rule of thumb has always been to think big and start small. Small is always doable, and doable is always good. A wise company will pick its own starting position on the sustainable track and manage its conversion to sustainability at a pace that accounts for its growth, capitalization, and other circumstances. Some will be able to move faster than others. But as long as everyone's moving, we'll get there. Like a puzzle, each piece placed will facilitate the faster placement of those that remain.

If it seems like all these factors add up to a challenge, you're right. The world changes constantly, and today's right decision can become tomorrow's wrong move. Building a sustainable system of commerce isn't going to happen overnight. But the handwriting is on the wall.

You can see it at the Home Depot, which recently agreed to cease selling old-growth redwood lumber products. You'll find it in the aisles of retailers like Wal-Mart and Ikea, which have now stopped sourcing products from Burma, a country with an appalling human rights record. It's in the blueprints for the Ford Motor Company's mammoth Rouge River complex, which is being rebuilt to address sustainable manufacturing and social concerns.

From Tweezerman, a manufacturer that offers employees schedules that accommodate their families' needs, to the compost pile behind your local diner, a new view of what business can and should be doing is emerging as companies large and small realize that we cannot continue indefinitely as we have. If we know this, and I think the point is fast becoming obvious even to those in deepest denial, then there exists no reason to delay. Make no mistake. The change is coming, and it will simply force itself upon us in the name of survival if we sit idly by during its approach. If we're wise, we'll avoid such an eleventh-hour response and its reactionary dangers and make the change while time remains to make it thoughtfully. We'll realize as a business community and as a society that the question really isn't why we should do it. The question is why on earth we would do anything else.

WHAT'S MONEY GOT TO DO WITH IT?

Vicki Robin

We are seeking a new American *dream*. Not a new American policy, a new American study on quality of life, or a new American statistic. Policies, studies, and statistics may help dreams come true, but dreams are the *real* drivers of change. So I begin with a dream of happiness—the true bottom line of all.

It's not too hard to imagine a simple life, richly lived. There would be ample time for friends and family. There would be spaces and pauses amidst the onrush of daily life to think, to wonder, to play with children, and to listen for the possible in the prosaic. There would be good work—and healthy leisure. The simple life. Just the term pulls some deep longing out of our bodies, down below the noisy mind or even the restless heart.

It's not even too hard to imagine a society that would support such a life for its members. There would be laws that encourage a living wage, protection of green space, eco-efficient products and buildings, European-style work years with ample vacations and lots of time off, enough competition for innovation but enough cooperation for everyone to feel safe. We would be diverse, yet working together. Some would get rich—but not so rich or isolated as to leave the rest of us behind. It would be a society where the best in us has space to grow and the worst in us has space to heal. Adventure would come both at the frontiers of science and the frontiers of soul. Commerce would serve real human needs and markets would be lively because humans will always have needs—and will enjoy the novelty and gossip of the public square.

We deserve to have this dream. Countless sustainability advocates preach it from the modern pulpits of conference halls around the globe. So what's stopping us from realizing it? Is it a lack of collective will? Is it bribery in politics? Is it that we are too tired at the end of the day to care? Or is it that we are so rich that we think we don't need to invest in the common good, preferring private pleasures?

I believe that how we look at money is a prime suspect in the mystery of where the good life has gone. In the inquiry into the chasm between our longing for a new American dream and our capacity to create it, we are suffering from three levels of money-blindness. First, we conflate means and ends. We confuse money—pieces of paper and metal that facilitate complex exchanges of goods and services—with nonmaterial needs and desires. Second, we have lost our imagination about how we might meet many of these deeper needs without money. Third, we are ignorant about the mechanisms by which the money economy actually aids and abets the cancerous growth of consumption globally.

By looking at these three confusions we will come to see that "enoughness"—a stance of material sufficiency and spiritual affluence—describes a transformative way of living that liberates humans to live in wholeness and balance. When we are blind to the truth of money, we glimpse "enoughness" and think we see deprivation. When we open our eyes to the hidden costs of money—to ourselves and life—we become inspired generators of human wealth.

The unexamined life may not be worth living, but the unexamined relationship with money is lethal to humans and other living things.

CONFUSION OF MEANS AND ENDS, OF MONEY AND HAPPINESS: WHAT *IS* MONEY?

In one sense, money is an agreement. We agree that little pieces of paper and metal have value, so we are willing to part with true wealth—time, skills, products, edibles, and the like—to have money. In Economics 101 we learn that this agreement called money serves as "a store of value" and "a means of exchange." It allows us to trade wealth today, or to save for future purchases. It is an artifact that exists to facilitate trade—no more, no less. If this definition were the last word and the whole truth about money, however, we would function far more rationally. Once we had the basics—food, clothing, shelter—we might devote ourselves to a raft of activities having nothing to do with money.

This isn't hopeless idealism—the few remaining indigenous peoples do live this way, even today. In fact, for most of our tenure on this earth—truly, millions of years—we didn't use money. Even as recently as a few hundred years ago in Western economies, only a fraction of the economy was monetized. Humans met far more of their material, social, and spiritual needs through their own efforts and through "reciprocity." The history of the West has been, in part, the history of separating from our ability to meet our needs directly or through community. It's been a story of migration, trade, resource wars, specialization, industrialization—all of which have increased our dependence on trading chits (money) to meet the majority of our needs. Money has become an end and not just a means.

One reason we cannot get "back to the garden of simplicity"—to that vision we all hold of a balanced life—is that we have forgotten that money is just a tool we invented. We now believe that without money we won't survive, when in fact we can't eat it, wear it, drink it, go anywhere in it, or meet any real needs with the pieces of paper and metal. Despite this, we believe that every morsel we chew, every stitch we put on, every swallow we take, every journey we set out on, and every pleasure we enjoy will somehow require money.

Because we believe that money is necessary for the fulfillment of our needs, we think we must devote the majority of our waking hours to earning it. This fixation on money as the key to getting what we want often displaces our other needs—for affection, security, intellectual stimulation, contribution to society, rest, play, creativity, and freedom.

And what is the result of this confusion of means and ends? Many of us engage in competitive consumption to satisfy ourselves that we are good enough, smart enough, successful enough, and maybe even a little better than others. Millions of us accept the spiraling costs of essentials like health care, education, and housing, not realizing that our measure of what is enough escalates as more and more people have what we have.

Indeed, most people in North America engage in the dominant myth of "more is better" without question and even good, caring people rationalize excess as necessity. Because of this, so many people, with a helpless shrug, say they need ever-more money to meet the demands of "modern life," citing a vague boogeyperson called "cost of living." "More is better" now means "more *money* is better." This acquiescence to excess then requires putting up thicker and thicker walls between our consciences and the billions of people who live in poverty. If we were selling our time—and perhaps our souls—to a system that truly fed us, that would be one thing. But the economy is not designed for people; rather, people are trained to serve the economy. In a downturn, the economy sheds people to preserve property and profits. And our relationship with money is intimately tied up with how dependent we are on this economy.

But what's the alternative? We all need money to survive. For money, we need jobs. For jobs, we need a thriving economy. For a thriving economy we need to serve the economy in some ways. Should we go back to the woods and eat nuts and berries? Deprive ourselves of the necessities of life in this complex world—phones, computers, cars, televisions, not to speak of houses, a hot shower in the morning, and good food? No.

One route out of this dilemma is to redefine money in terms of something real to us, rather than abstracted access to a never-ending stream of goods and services. I offer this alternative to "store of value" and "means of exchange": Money equals our life energy. By this I mean the hours we invest on the job to earn it. Time is all we tangibly have on this earth. In our youth it seems unlimited, but somewhere just south of forty or so we become aware that our days are numbered. Every hour we invest on the job is an hour *not* invested directly in our children, our mates, our community, our health, our spiritual development, our search for meaning, or our contribution to the larger life. Our jobs may, in more or less abstract ways, relate to all these other spheres. We earn money to support our family. Our professions are sometimes essential to the social fabric. We learn lessons through our jobs that imprint on our souls.

In *Your Money or Your Life,* which I coauthored with Joe Dominguez, we offer a simple calculation to understand how many pounds of flesh one invests each year for the dollars in one's wallet. Through this calculation, readers see that their nominal salary, say twenty dollars an hour, is not their real hourly wage. That amount must be adjusted by all the "unpaid" hours they invest in their job (commuting, worrying, dressing, recovering, vacating) and all the money they spend to maintain their job (transportation, clothing, lunches, child care, education, entertainment—not to mention taxes). Twenty dollars minus expenses can quickly shrink to five. A sixty-dollar dinner, then, isn't a reward for a tough day on the job; it's more hours in the workplace.

With this understanding in place, people learn to track their spending, tally their expenses each month in categories that reflect their particular habits and convert each total from dollars to "hours of life energy." Then, for every category of purchase, they ask three questions:

1. Did I get fulfillment in proportion to the hours of life energy I invested in this category?

2. Is this expenditure of my life energy in alignment with my values and life purpose? This question ties everyday economic choices to deeper yearnings for love, community, and meaning.

3. How might this expenditure change if I didn't have to work for a living? This question helps people distinguish between basic needs and job-related ones.

To encourage answering these questions honestly, we suggest the mantra: "No shame—no blame."

By going through this process people quickly realize—without admonishment or self-deprivation—the profound cost of money to their well-being. They see that they can have the nonmaterial benefits of consumption—happiness, security, self-esteem, and the like —for little or even no money, but that they can never retrieve the hours they invest in getting the money. On average, this simple insight, verified through tracking and evaluating their expenses, leads to a rapid drop in spending by, on average, 20 percent. The veils drop from their eyes. Efforts to make them spend money seem ludicrous. They brag about the clever ways they meet their needs for much less money. Instead of confusion or guilt over their partici-

pation in American excess, people find that being mindful with
money is:

- liberating (they often *don't* buy things they used to buy)

- fulfilling (they consciously enjoy what they spend their "life
  energy" on)

- meaningful (every transaction is a chance to express their values)

On the other hand, some perennial penny-pinchers find themselves spending more to achieve that sense of "enoughness." The
game gets even more interesting as people realize that they can
support ethical merchants or companies by buying their products—
or investing in their stock. Over time, as their expenses decrease
and debt melts away, savings appear, which can be invested for income, spent on possessions that will last a lifetime, or donated to
further reinforce their values.

People begin to live within their means. Competitive consumption and competitive earning both diminish. They have financial intelligence—they know where their money is going and are wise consumers. They have financial integrity—they make values-based
choices about spending. And they have financial independence—
they are ever-more liberated from the dictates of the money system.

As people realize that the flow of money through their lives and
through the economy is related to both ecological health and social
equity, they understand that mindfulness with money is one of the
most powerful ways to express care for the earth. In fact, they
slowly evolve into a lifestyle that—for all the pressure of the consumer culture—looks surprisingly like the ideal described in the
opening paragraphs of this chapter.

Is this easy? No, not necessarily. It takes focus and often support, given that there is little reinforcement for frugality in the general culture. But millions of people who have discovered the vast
middle ground called "enoughness" live among us. We've studied
them—albeit informally—and here are our findings.

People who know how much is enough have everything they

want and need to live a life *defined by themselves* as fulfilling and meaningful. They have a purpose for their lives larger than simply meeting their individual needs. They swim in a bigger pond, and all their choices—from how they spend their time to how they spend their money to who they hang out with—reflects that chosen context for their existence. Very often, just asking the three questions monthly reveals this need for a larger purpose and spurs a dawning realization about what that might be.

People who recognize sufficiency also have an internal yardstick for fulfillment. Their "enoughness" comes from honest self-examination. They don't compare their assortment of stuff to someone else's stockpile to assess whether they indeed have enough. In a "more-is-better" world, this steadfast gratitude and authenticity is quite a feat—but surely both possible and worthy of devoted practice.

We've also noticed people who have a sense of "enoughness" develop economic resilience through learning an ever-expanding set of skills. It might start with cooking from scratch. Or bicycling to work—and then learning to fix the bike. Smart shopping comes naturally—buying in bulk, learning to comparison shop. Yard work substitutes for the gym. Or maybe lawns become vegetable gardens. Baby-sitting cooperatives allow parents to save money *and* have a life. Greeting cards are made, not bought, using paints or pressed flowers. Music gets made, not bought. And on and on.

Linked to this skill set is often an expanding network of mutual aid. Neighborhoods organize informal barter systems for everything from respite for caregivers to wheelbarrows. Other people are no longer seen as a nuisance or a threat. Magically they transform—in the eye of the beholder and barter partner—into marvelous, gifted, and talented beings. In times of threat—be it a terrorist attack or an earthquake—people naturally turn to one another for solace and security. They want to know what skills, supplies, and services are available within walking distance. They want to know in what ways they are dependent on larger systems and what they can do to provide for themselves. These are the questions and concerns that come naturally to people who focus on balancing the material and nonmaterial sides of life. But they don't need disaster to remember to value human connections and to share.

In 1990, as word of the tape course that preceded *Your Money or Your Life* spread, a reporter, Sarah Goodman, came to my home to do the first of now thousands of media interviews.

I dressed carefully. It's one thing to be frugal. It's another to look it. Sarah chatted me into comfort as good reporters do, approaching again and again the question of how I lived on $600 (now $900—thanks to "health" insurance and other costs of growing older) a month. My explanations were all common sense. I'll recite some of them here, with my usual caveat that this is *my* "enoughness," not my recommendation of how others should live.

Shelter

I co-own a large house in Seattle overlooking the Cascade Mountains where I live with four other adults (and a steady stream of guests). I have a small room of my own and share large common spaces. Some would call it crowded. I call it convivial. I cook less but eat better. I clean less but live in an orderly and beautiful home. I do less yard work but have more gardens. I spend less and am never lonely.

Clothes

I happen to love thrift stores. Also, the more I know who I am, thrift-store variety actually feeds my imagination—I've shifted from fashion to style.

Food

Within reason we eat food that is homegrown, in season, and local so our food tends to cost less.

Travel

I *do* love to travel! Fortunately, my life as a public speaker means other people are willing to pay my airfare. I only sacrifice control over where I go and when. I've made it to places like South America and India and Europe and Thailand—but not on my time-table and not on my dime.

Insurance

The minimum! High deductibles plus taking care of what I have (car, house, *and* body) plus a philosophical attitude allow me to feel adequately secure. Even so, this is my biggest luxury—almost a quarter of my expenses are for insurance!

Fun

"But what do you do for fun?!" reporters always ask. (Subtext: frugality is boring and limiting.) A large part of the answer is that my joy is an inside job—I simply take pleasure in living. My commu-nity service (which I call "strategic meddling") keeps my creativity and compassion flowing. And I do like to take walks (especially with friends), read books, watch videos, and have conversations over coffee or a meal.

Services

When Sarah asked about the costs of paying others to fix and maintain things, I said . . . "I barter. When I have a physical problem, my nurse friend tells me if it is anything to worry about before I commit time and money on a doctor. Most often, the problem is easily explained and self-limiting. In 'exchange,' I cut her hair when-ever she needs it."

"And doesn't she mind?" Sarah asked.

"No, why?" I really didn't get it.

"Well, she's a professional. Doesn't she mind not getting paid?"

"What?" I answered without thinking. "I charge her $25 for the haircut and she charges me $25 for the consult and we'd each pay Uncle Sam $10 for the privilege of using his money?"

We blinked at one another in silence for several seconds, as though we were each on a tour bus observing the odd natives of another land.

MORE THAN MONEY—MEETING NEEDS DIFFERENTLY

One of the things money obscures is that we often meet our needs without it. With creativity, the proportion of money-based need fulfillment can shrink even further. Ask yourself: Which needs can *only* be filled by money? Primarily, taxes for public services and trade require "real" money. In fact, there are actually two kinds of money: cash for the outside world and barter chits. Some needs *need* money. Some needs are met through personal skills and social networks. And some needs—fun, play, love, imagination, meaning—are met nonmaterially.

Hazel Henderson coined the term "the love economy" to refer to the approximately 50 percent of all transactions globally that meet needs without money exchanging hands. The services we provide for one another within the family are the love economy in action. Most elder care—a growth industry in the aging boomer United States—happens elsewhere in the world through the love economy. Neighborliness—dog walking, tool sharing—is part of the love economy. Carpooling is at least the "trust" economy. While "love" may have nothing to do with it, most transactions between departments in corporations do not show up as monetary trades; the recipient of a shipment of parts from warehouse to factory floor "signs for it." In fact, we are quite accustomed to informal trades with people we accept as part of our family, club, clan, or workplace. The left hand and the right hand *do* share.

Many of the ways we "save money" are also part of this nonmonetary economy. Air miles are actually an alternative currency; they aren't dollars. Grocery coupons are another form of alternative currency. Reduced-cost matinees or luncheon buffets are ways that proprietors sell the same product for less money. Clearance sales are the same principle—25 percent real money, 75 percent alternative currency (this time no coupon needed) and you get the same product.

Barter itself is making a comeback—and it's far more sophisticated than trading a pig for a bushel of tomatoes.

In Curitiba, Brazil, for example, slum dwellers can trade their separated garbage, right in their community, for fresh vegetables, bus tokens, or school supplies, thus saving the city garbage-collection costs and providing poor people with essentials. In Minneapolis the Commonweal Community HeroCard is like a credit card but one that draws from both a money bank account and a volunteer hour bank account as well. It's called a "smartcard"— that's very smart indeed. Businesses that might be willing to take Commonweal Community C$D (Community Service Dollars) in partial payment for underutilized resources (remember those matinees . . .) join the program. At the same time, participating community service organizations join and are issued C$D with which they can "pay" their volunteers, who then spend those C$D with participating merchants, who then donate a percentage of their profits from those sales back to the Commonweal Community. Certainly, there is much more to this system, but this short description points to a formalized, community-wide way that complementary and normal currencies can exist side by side.

DARKER CURRENTS OF OUR CURRENCY

These stories aren't just moving tales of a few inventive souls. They reveal that money as "barter chits" can be constructed in ways that foster community and conservation. The opposite tendency comes out of our current coin of the realm.

Few of us realize that our money is no longer backed by anything real—not by gold, not by land, not by heads of cattle. Nor do we realize that the government doesn't create and control money. Money comes into being when banks make new loans. This isn't money that existed in savings accounts—it comes into existence as a line of credit that must be paid off with interest. Fully 95 percent of money is based on interest-bearing bank loans.

What does this mean? Every time money is created, there is an interest cost. For example, say you buy a $100 item with a credit card and pay the minimum monthly balance. By the time you've paid off your bill, you've probably paid an additional $200 in interest—and the item is often broken, lost, or sold in a garage sale. Where did the $200 of interest come from? You had to work for it. Getting and keeping your job meant that others lost out in the competition—increasing unemployment and increasing the demand for job creation. Job creation means economic expansion. Economic expansion in a finite world means someone or something is bearing the cost—and most often it's the environment or the poor elsewhere. The same is true as businesses and governments borrow money to pay off past debt or to fund future expenses. This money didn't exist before the loan was made, and the loan must be paid with interest. Growth *must* occur or debilitating deficits will take the institution down. We consider this natural, but the natural world does not—and is no longer silent on the issue. The by-products of growth—pollution, resource depletion, global warming—are speaking ever-more loudly.

The artifact we use to acquire our daily bread is driving us into the very environmental and social destruction we so desperately want to avoid. How money is created, by whom, and on what terms actually creates an artificial world of winners and losers. One of the most famous movie lines of the 1980s came from Gordon Gekko in *Wall Street:* "Greed, for lack of a better word, is good. Greed is right. Greed works."

Greed may work, but so does sharing—both small scale and more systemic. It is surprisingly simple to have all we need through more frugal and equitable means. "Public consumption"—providing shared resources for the whole community—provides every one of us with a higher quality of life. While few of us consciously link the pain of taxes with the pleasure of functioning communities, taxes do buy books as well as fund libraries to store them so they are available for all to read. Our tax dollars buy roads and sewers and other community infrastructure—making these items one less necessity on our personal shopping list.

There are some people who simply cannot imagine further frugality. They don't have enough. These are the two billion living on one dollar a day, the low-income single parents who economize, pinch pennies, and still can't meet needs without greater income. A truly sustainable future requires systemic changes to address the genuine material and financial needs of the poor. But for many, frugality can be a path to liberation.

When you put together personal frugality with the vast array of tools and mechanisms for meeting one's needs through informal trade, social networks, and public consumption, the "work and spend" treadmill of individual consumption slows down enough to let a lot of people off. Money is not such a mystery. It might not even be the insurmountable barrier to having the life you want. There is wealth beyond money. Riches are interior and interpersonal as well as material. Assets include character and community as well as property. Knowing this, you are strong in ways that the economy can't touch. You have enough.

BE A LOCAL HERO: STRENGTHENING OUR COMMUNITIES, HEALTH, AND ENVIRONMENT BY EATING LOCAL

Mark Ritchie

Growing numbers of Americans are buying local. Whether it's tapping into old instincts or searching for new social outlets, we're doing it more, enjoying it more, and making a positive difference in the process. Over the past few years, there has been a slow but steady shift away from the old food economy, where "cheapest is best"—no matter what the cost to society. While we can buy strawberries in the middle of winter in Minnesota by air freighting them in from China or Chile, the costs to the society and to the environment are much, much greater than the price at the cash register.

More and more businesses and shoppers are gravitating toward a new food marketplace, where farmers and eaters are building linkages that make it possible for consumers to buy an increasing percentage of their food from small, local producers they know and trust. The steady annual growth in the number of farmers' markets, direct-selling farms, and u-pick operations are just a few of the indicators of consumers choosing to "vote local" with their food dollars. While many of us will continue to enjoy the benefits of the global market, such as coffee and chocolate, the movement toward a more local food system, especially for basic food necessities, deserves our personal and political support.

Part of the desire to buy local comes from the better taste, higher quality, and lower price that is inherent in most local, fresher foods. Another motivation comes from the increasing consumer awareness of the effects of the industrial food system. Unsafe foods, inhumane treatment of animals, unhealthy working conditions for farmworkers and farmers, contamination of the air, water and soil, and devastation of rural communities are among the devastating impacts of the system that currently dominates our supermarkets and restaurants.

While the trend toward localized food systems has been going on for quite some time, it has been greatly accelerated by several recent events. First, the downturn in the overall economy seems to be pushing even more consumers toward the savings that come from buying directly from producers at farmers' markets, in food cooperatives, or from community-supported agriculture (CSA) farms. Buying local makes it possible to bypass the middlemen and monopoly corporations that squeeze both farmers and consumers. This trend toward more buying at farmers' markets and other lower-cost options like u-picks and farm stands is consistent with experience from the last major recession in the early 1980s.

A second factor has been the growing public concern about the safety of our food, including new worries about bioterrorism since September 11. People want to know who grows their food and how it is produced. Buying local makes this information easier to find. You can visit or call the farmer, talk directly to the buyer at the food coop, or get important information from the market manager at the farmers' market. Local producers that are certified organic are following important food safety rules, such as avoiding hormones and antibiotics, which are designed to protect human health. Bioterrorism is a concern for consumers who purchase foods shipped long distances and across borders that are possible targets for organized crime or random terror.

A third factor is growing public awareness about the links between the environment and methods of producing and shipping food. A wide range of ecological concerns, from global warming and contamination of rivers by factory hog farms to the dead zone off the coast of Louisiana are increasingly understood as an unacceptable cost of the old food economy. In addition, the current system of transporting heavily chemicalized and artificially preserved food long distances has been shown to have terrible consequences for the planet, contributing to the destruction of the ozone layer, global warming, contamination of the air, ground, and water, and diminished biological diversity.

While each of these factors—cost, human safety, and the environment—are important public motivators, we know from numerous public opinion polls that the primary motivation for consumers to buy local is their desire to support their neighbors working on farms and other businesses in the local food chain. A recent survey of western Massachusetts residents mirrored what other pollsters find around the country: Local residents want to support their neighbor farmers whenever possible by buying their products.

Not surprisingly, the global corporations who benefit most from the old food economy are trying hard to reverse the trend toward local food systems. Their main weapon, of course, is money. Each year they spend billions of dollars on advertising to convince shoppers to ignore their desires to buy local. They spend millions to finance politicians who support laws that favor big business over small. One of their more effective strategies is using taxpayer funds to subsidize large-scale industrial units, such as the gigantic hog factories.

Global corporations also use their political influence to get massive subsidies for the long-distance food transportation system. Hundreds of millions of our taxpayer dollars go to subsidize the airfreight transport system in ways that significantly benefit large food corporations. At the same time, however, many states and counties do not have the funds to maintain the local roads and bridges needed to enable local producers to bring their products to market. The cost of gasoline and diesel fuel, a major expense for long-distance food shippers, is heavily subsidized in the United States, through special tax breaks and direct and indirect subsidies. While these subsidies are significant, they don't even include the unpaid costs of polluting our air and water, the long-term cost of altering the climate, or the incalculable price of our deep dependency on the Middle East for oil.

There are times when these subsidies for the corporate food system can make industrial food products that are shipped across the country cheaper than local items. There are also instances where the use of child, prison, or sweatshop labor can make an item shipped in from another country cheaper at the cash register than a local item. However, in many instances the cost of long-distance shipping makes imported items more expensive. And monopoly control leads to unjustified higher prices. Shoppers do care about prices, but they also care about convenience, quality, their families' health, and the environment. Local food systems are growing at such a rapid rate of speed precisely because they deliver the kinds of values that customers want—in both price and quality.

While the obstacles erected by food companies and some of their allies in government are significant, they have been unable to change the underlying market trends toward more local food systems. Hundreds of thousands of farmers, market gardeners, and independent food companies are now responding to signals from millions of consumers who want to buy food from someone they know and trust.

This chapter explores five regions where resurgent local food systems give hope and direction for a truly sustainable future. These new food economies are improving the lives of urban, suburban, and rural residents by restoring fairness to all actors in the food chain—from producers and processors to food industry workers and consumers—and by changing food production, processing, and distribution practices to make them as environmentally sound and health-promoting as possible.

RENEWING THE COUNTRYSIDE—EXAMPLES FROM AROUND THE WORLD

Southeastern Ohio

On the edges of Appalachia in southern Ohio lies Athens, a college town in the midst of one of the poorest regions in the nation. In Athens and its surroundings a number of organizations and projects have created one of the most exciting and dynamic local food systems in the country.

One of those organizations is the Appalachian Center for Economic Networks (ACEnet). ACEnet started in 1986 as a community kitchen and has developed into a comprehensive approach to adding value to local products. June Holley, a former farmworker and inner-city teacher, created ACEnet based on models of relationship-centered economic development that she had studied in northern Italy. Holley believed that the low-income, subsistence Appalachian farmers in her region could begin combining their products and individual skills in ways that added value to each other's products and eventually created an economic niche for the entire region.

With foundation and church support, June, Leslie Schaller, and other ACEnet staff converted an old lumberyard into a bustling center that includes the kitchen and other food preparation facilities, a retail store for test marketing new products, and a huge new cooking facility for manufacturing specialty food items. There are also meeting rooms, a library full of food safety information and government regulations, and an Internet-based research center. Fifteen years later, ACEnet has proven to be one of the most successful incubators for food business entrepreneurs in the entire country. It has spawned more than 120 food businesses, created several hundred jobs, and contributed to a whole spectrum of projects in more than a dozen southeastern Ohio counties that are working to transform their entire rural economy. These include the Athens Farmers' Market, a thirty-year-old operation drawing more than one hundred farmers yearround; Rural Action, a food project called Good Food Direct, a sustainable forestry initiative, and the National Center for Preservation of Medicinal Herbs (NCPMH).

The NCPMH is part of a unique green corridor dedicated to preserving the high biodiversity of this unique region. This sanctuary corridor, which now includes over two thousand acres, started nearly thirty years ago when nineteen-year-old Paul Strauss moved from New York to an eighty-acre farm and started a small herb company call Equinox Botanicals. Over the years Paul's efforts blossomed into a unique national treasure of biodiversity and innovation. The NCPMH and their neighbors have created a working landscape and forest where herbs that are threatened by overharvesting are being cultivated and studied in a living laboratory.

This region is a good example of how the local, the national, and the international can fit together nicely to support a vibrant local economy. While most of the products come from local producers and are processed right there in Athens, a significant percentage of the medicinal herbs and other locally produced goods are sold nationwide or even exported to specialty markets overseas. By putting an emphasis on the sustainability of the production process—

meaning that the farmland and forests that are producing the raw materials for these products are being managed to ensure permanent productivity—local groups in southeastern Ohio are working to ensure that the supply and demand equation equals a high-quality of life for both present and future residents of this region—not just short-term gain at the expense of future generations.

Vancouver, British Columbia

Vancouver's local food economy is hot. An organization called Farm Folk/City Folk and a contingency of demanding chefs in search of fresh, local organic produce are two of the forces fueling Vancouver's burgeoning local food scene. The city now boasts almost two dozen home delivery organic food suppliers, as well as ProOrganics, one of the fastest growing organic food distributors in North America.

Driven by founder and director Herb Barbolet, Farm Folk/City Folk has spent much of the last ten years pulling together different constituencies in the local food system (the food, farming, health, and environmental communities) into a cohesive force to shape new public policies in support of local food production. They have won important victories in the areas of farmland production, government procurement of locally produced foods, and adoption of local food policies by area governments.

This magnificent fabric of rural and urban cooperation is displayed each fall at Farm Folk/City Folks' Annual Harvest Festival, a celebration of local foods, wine, and music. Hundreds of local food producers and tens of thousands of consumers of all ages gather at the festival for a glorious celebration of their local food system. This work is supported by a remarkable program based at the University of British Columbia (UBC). Moura Quayle, dean of the faculty of agricultural sciences, has transformed UBC's fairly conventional school of agriculture into one focused on sustainability. At UBC, faculty, staff, and students have begun reviving and redesigning the campus farm to provide leadership in agro-ecological design, community planning, and development. Students are developing a market garden that provides food for students, faculty, staff, and community members, as well as education and volunteer programs that provide on-farm applied learning opportunities and build strong community connections.

Another local food economy project in Vancouver is the Eco-Cafe Sustainability Society. Their flagship is the new EcoCafe Sustainability Centre, created to help young people develop their capacities to create ecological systems by using sustainable technologies.

Canada's Office of Urban Agriculture, called City Farmer, works closely with many of these projects and has promoted urban food production and environmental conservation in Vancouver for nearly fifteen years. City Farmer runs a number of demonstration urban food projects, including a garden in a nearby residential neighborhood. The organization also built the world's largest online archive of urban agriculture information available through its website at www.cityfarmer.org.

On the public policy side, the Vancouver Food Policy Organization works to enhance food security by providing a sustainable local food supply that is accessible to all. Created in 1994 by a group of community nutritionists, the organization has been the leading voice on using the region's local food production capacity to feed the hungry and malnourished in Vancouver. They created a hunger hotline to connect those in need with the nearest free/low-cost food resources and produced an analysis of ways to increase use of local food projects in school meal programs. Other work includes public education, local fruit gleaning projects, and development of community food security indicators.

One of Vancouver's most innovative local food security initiatives is the Good Food Box, which helps people in need obtain quality, fresh, and nutritious fruits and vegetables. The program does not involve handouts. Rather, the Good Food Box uses centralized purchasing and economies of scale to provide its participants with yearround access to high-quality food at the lowest possible prices. The program promotes the self-reliance and dignity of its participants by permitting them to purchase food and to be directly involved in the program's operation. Each month, more than one thousand participants assume an active role in helping themselves and their children to better diets and, ultimately, to better health.

The Vancouver Community Kitchen Project assists residents in building their own local food economy to ensure food security for hungry and low-income families. Its unique mission is to create opportunities for people to cook together through the community kitchen movement in Vancouver. The project offers technical assistance, written materials, community contacts, and practical advice to anyone interested in starting a community kitchen; provides ongoing support to more than forty existing kitchens in the Vancouver area; and works to build active relationships with other food-security projects.

These examples are only a few of the incredible community-based companies and organizations that have successfully created North America's largest and most dynamic local food system. These new initiatives are slowly developing as a serious alternative to British Columbia's old economy, which is hopelessly dependent on unsustainable, destructive, and old-fashioned industrial models of mining, forestry, fish farming, and agriculture.

The Netherlands

Over the past fifty years, the Netherlands has followed two dramatically different paths in industrialized agriculture. After the Second World War, the Netherlands embraced industrialized agriculture full force and took it further than any other country. They had the "benefit" of very bountiful natural gas supplies, which encouraged the country to rely heavily on artificial fertilizers, heated greenhouses, and other energy intensive methods of industrial food production. It was a spectacular success in terms of productivity but an unparalleled disaster for the environment.

In recent years, however, the negative social, ecological, and economic impacts of this method have become so painfully clear that they can no longer be ignored. Dutch farmers and consumers have sought alternatives to industrialized farming and are now building the most extensive New Food Economy in all of Europe, perhaps on the planet.

What caused such a dramatic shift? The change occurred in three places—on the farm, in the marketplace, and in the broader society. On many farms, the impacts of the industrial model became dramatic. Mass animal disease epidemics, like swine fever that necessitated the slaughter of almost all of the pigs in the Netherlands, foot and mouth, and mad cow disease exposed the cruelty and unsustainability of the industrial livestock-rearing model. At the same time, chemical contamination of much of the water from the runoff of pesticides and other chemicals nearly wiped out a number of greatly loved species of birds and other animals. At more or less the same time, consumers began to signal the need for radical change. German consumers began complaining that Dutch tomatoes, although beautiful, were tasteless. Then Danes complained about the poor quality of the famous Dutch cheeses. Obsessed with quantity, the Dutch found that diminished food quality was eroding their market share. The third factor was the Dutch public, who began to complain about the industrial model's increasingly negative impacts on their water, nature, and landscapes. By this point, farmers and their marketing cooperatives began to understand that old ways of producing and distributing foods were no longer working—they were not generating enough profit to enable families to stay on their farms and they were driving away customers and upsetting the public.

Always innovative, Dutch farmers sat down with consumers, environmentalists, scientists, and others to talk about all aspects of this economic, social, and ecological agricultural crisis. Out of these conversations arose a change in orientation, and a new approach began to develop. First, farmers took tremendous strides toward improving the quality of their food and returning to values of freshness, taste, regional differentiation, and safety. Second, they began adapting their farming practices toward methods that were softer on nature and protected the environment. As producers switched to more organic and sustainable production methods, environmental groups such as Friends of the Earth went from being harsh critics to strong supporters. Third, a number of farmers began looking beyond simple production of food products toward a more diverse and secure livelihood. They focused on ensuring their economic security by making innovative links that provide on-farm revenue while advancing environmental protection.

Farmers joined with bird-watching groups to increase habitat and bird populations; they partnered with drinking water companies to provide low-cost water filtration and protection services via their farmlands; and they contracted with local and regional environmental agencies to increase wildlife and restore endangered species. New forms of green tourism were born that brought together the city and countryside in mutually beneficial ways, providing a new revenue source for farmers and excellent, inexpensive holidays for urban residents.

Most recently, the Dutch Farmers Union, in collaboration with consumer and environmental groups, launched the Green Virtual Supermarket. Locally grown and sustainably produced foods can be ordered online and delivered fresh each day. Farmers' markets and other direct-buying opportunities are springing up all over. Dutch farmers have reclaimed their reputation for high-quality foods, and consumers have reclaimed their connections to farmers.

Kerala, India

The southern state of Kerala is one of the cash-poorest regions of India, but (under a communist government!) through state policy it has become one of the richest and most socially wealthy regions in the world. In addition to a nearly 100 percent literacy rate and an extensive system of social benefits, Kerala has done a remarkable job of feeding its population of more than thirty million people through local food systems. Although the per capita income is one of the lowest in India, around $350 dollars per year, the life expectancy, birth rate, and overall quality of life rivals some of the richest nations on the planet.

Although much of this development has come in recent years, there are traces of democratic life that fueled these remarkable achievements going back to the nineteenth century, despite British colonization. As author Bill McKibben has observed, "Kerala suggests a way out of two problems simultaneously—not only the classic development goal of more food in bellies and more shoes on feet, but also the emerging, equally essential tasks of living lightly on earth, using fewer resources, creating less waste."

A good example of Kerala's resourcefulness is the collaboration between the local government and residents to launch the People's Resource Mapping Program, which spreads "land literacy" throughout the state. Villagers created maps of their regions showing water tables, topology, and land use patterns, to better understand and manage their resources, reduce erosion, and boost food production. In one village, residents reduced their dependency on expensive imported vegetables by using the maps to identify patches of land that were available for market gardening. As a result, 2,500 young people were employed to create gardens and sell vegetables in local markets at prices much lower than the imported products.

As McKibben describes, "This is the direct opposite of a global market. It is exquisitely local—it demands democracy, literacy, participation, and cooperation. The new vegetables represent 'economic growth' of a sort that does much good and no harm. The number of rupees consumed, and hence the liters of oil spent packaging and shipping and advertising go down, not up."

Kerala also has focused on the wider issues of healthy households, creating a network of clinics and medical care facilities that are unparalleled in poor countries. As part of this effort, all mothers are taught breast-feeding and there is a state-supported nutrition program for pregnant and nursing mothers. As a result, infant mortality is nearly the same as the United States.

There are many challenges that face Kerala going forward, as the global economy reaches further into their region. For example, some of their local food-processing industries, such as cashew and coconut fiber processing, are relocating to areas with much lower wages and fewer environmental regulations. At the same time, many of the highly educated youth are being attracted away to work in other parts of India and in other countries. Local leaders are aware of these problems and are working to find new, sustainable strategies for both creating new industries and keeping their young people involved in the region.

In the past few decades, many people have traveled to Kerala to see their successes firsthand, hoping to shape the development process in their home countries of Vietnam, Cambodia, and Peru.

Increasingly, it is clear that some of the lessons from Kerala are also important to rich countries such as the United States, where the challenge is to reduce consumption of material goods in order to improve the quality of life. Again McKibben: "Kerala does not tell us precisely how to remake the world. But it does shake up our sense of what's obvious, and it offers a pair of messages to the First World. One is that sharing works. Redistribution has made Kerala a decent place to live, even without much economic growth. The second and even more important lesson is that some of our fears about simpler living are unjustified. It is not a choice between suburban America and dying at thirty-five, between agribusiness and starvation, between 150 channels of television and ignorance. It is a subversive reality, that stagnant/stable economy that serves its people well, and in some ways it is a scary one. Kerala implies that there is a point where rich and poor might meet and share a decent life, and surely it offers new data for a critical question of our age: How much is enough?"

Minnesota

Deep in the heart of corn and soybean monocultures, consumers in the metropolitan region surrounding the Twin Cities of Minneapolis and St. Paul are partnering with farmers and market gardeners to build a dynamic local food system. As with most other initiatives, the twin goals are to bring social and economic well-being back to rural communities, while ensuring fresh, nutritious, and delicious food at a reasonable price for consumers.

Out of the depths of the Great Depression, Minnesota's hard-pressed farmers and workers turned to cooperatives in the 1930s in hopes that these economic institutions could buffer them from the injustices of the capitalist business cycle. A Farmer-Labor Party was built based on cooperative principles, and the Twin Cities became a national center for cooperatives of all kinds. The Twin Cities region now boasts more than a dozen large and an equal number of small stores, along with cooperative wholesalers, bakeries, and other supporting institutions. The coops not only provide high-quality, reasonably priced organic and sustainable foods, they also help to fund a wide range of local food and social institutions.

Thanks to this well-organized wholesale and retail distribution network, farmers and small-food processors are able to market much of their production locally through the cooperatives. Youth-run farms, farmers' markets, and many community-supported agriculture farms on the edge of the metropolitan area have also sprung up as a result. One of the most exciting new elements is the dramatic increase in the number of small farms operated by new farmers who are recent immigrants from Cambodia, Somalia, Ethiopia, Mexico, and Central America. This has in turn created a new local restaurant scene featuring ethnic cooking from locally produced products.

Expanding out from the Twin Cities, this growing local food economy is spawning a movement to build local food systems in a number of other regions of the state. With some government funding, citizen groups are building regional food networks, including a significant number of small processors and distributors. Perhaps the most advanced of these is in the southeast region, where a special public-private partnership called the Experiment in Rural Cooperation has been spearheading a powerful coalition of local bankers, businesses, mayors, educators, and farmers that is creating one of the most exciting new initiatives in the country. Many of the most innovative of these projects and businesses are highlighted in a beautiful and inspirational book called *Renewing the Countryside,* which was published last year by the Institute for Agriculture and Trade Policy. A website devoted to promoting Minnesota's regional food system can be found at www.mncountryside.org.

BARRIERS TO BUILDING SUSTAINABLE LOCAL FOOD SYSTEMS

While great strides are being made, a number of laws and policies make it difficult, and sometimes illegal, to move toward a better system such as organic sustainable farming. For example, recent trade negotiations that have led to regional and global free-trade deals such as the North American Free Trade Agreement and the World Trade Organization have put highly restrictive conditions on the use of consumer labeling to promote locally oriented sustainable food systems. At the same time, the United States Department of Agriculture has shifted the focus of research funding away from ecologically sound pest control practices such as biological controls and integrated pest management and toward dangerous new technologies, including the insertion of pesticides in our food. The USDA's aggressive reduction in research on pest control without dangerous chemicals and pesticides makes it harder for organic and sustainable farmers.

One new technology the USDA is promoting is the use of genetic engineering to inject the pesticide *Bacillus thirengensisa* (Bt) inside corn, potatoes, and other crops. Bt, a relatively safe pesticide, has long been one of the few natural pesticides approved for use by organic, sustainable farmers under the strict rules for organic certification and labeling. When Bt is overused, however, the bugs become resistant, and farmers can no longer control them with low-dosage pesticide applications. Many organic farmers have suffered significant financial loss due to pest losses because Bt no longer works. For a few, this has meant the loss of the farm itself. To top it all off, genetically engineered Bt crops seem to be responsible for some of the massive killing of "good" insects that act as pollinators, such as bees and butterflies, which leads to lower yields in crops.

Another way the USDA has attacked organic farmers is by attempting to redefine the term "organic" so factory farms can easily meet weak standards. If the USDA is allowed to do this, it will flood the market with foods that are labeled "organic" but contain genetically engineered products and are produced with food irradiation and toxic sludge. Consumer confidence in organic farming may evaporate if the USDA is successful in pushing this agribusiness agenda into national law.

Perhaps the ultimate insult of the current system is the way all farmers and food system players are lumped together with agribusiness. All food becomes suspect when some is tainted and unsafe. All farmers get blamed for the pollution caused by a few. Consumers cannot easily distinguish or discriminate between the product of careful stewards of the land and those who are destroying the planet. Even something as minor as allowing farmers who choose not to use artificial growth hormones in milk to label their products as "hormone free" was prohibited in several states and is still discouraged by the federal government.

As family farms are wiped out, they are replaced by global monopoly corporations that attempt to control the food chain from pitchfork to dinner fork. Weak U.S. antitrust laws and lack of enforcement of current laws allow these monopolies to continue to expand their control. Unless we find ways to make it possible for family farmers to stay on their land and farm in the right way, we will find that these deadly but legal policies have become the norm, not the exception.

BE A LOCAL HERO

Large food corporations will do everything within their power to persuade society that their approach—chemical, energy, and capital-intensive production—is best. Television ads arguing that we need genetically engineered foods to feed starving children in Africa are the latest salvo in their public relations campaign. However, many people see through the corporate propaganda and are trying to persuade society that we need to remedy our approach to be able to feed ourselves. A new food economy is developing, out of the quiet work of local heroes who are producing and selling us fresh, wholesome, safe, and nutritious food grown nearby. And we can all take part in it—with each dollar we spend on locally grown food and each seed that we plant in our own backyard.

SPRAWL: THE AUTOMOBILE AND AFFORDING THE AMERICAN DREAM

Hank Dittmar

In 1998, there were 184,980,187 licensed drivers in the United States, and 207,048,193 licensed motor vehicles. For many decades, transportation analysts warned of the economic impact of the saturation of the domestic auto market as the ratio of drivers to cars neared one to one. But the nation reached the supposed saturation point and kept going. Now we view cars as lifestyle objects and need multiple vehicles for our many faceted lives—or our many faceted fantasy lives. In 1995, 91 percent of all households owned at least one car, and 59 percent of all households owned at least two cars. The wealthier a household is, the more vehicles it owns. In 1995, households with incomes under $20,000 owned an average of 1.3 vehicles, while households with incomes over $80,000 owned 2.4. Rising motor vehicle ownership is also encouraged by government policies. The federal government has subsidized the construction of roads and highways, kept gas prices low, and, with pressure from special interest lobbies, greatly limited the availability of alternative modes of transportation. In 2000, *Ad Age* magazine estimates that the seven largest automobile manufacturers spent over $11.9 billion on advertising for new cars, while the United States government invested only $7 billion on mass transit systems. As a result of relentless automobile advertising and equally persistent assaults on public transit by right-wing groups opposed to all things remotely public, auto ownership is associated with wealth, style, physical prowess, daring, and sexual conquest, while transit use, rail, biking, and walking are seen as either public bailout systems or dangerous and degrading activities. I find it curious that we subsidize the auto industry to the tune of billions of dollars while public investment in high-speed rail, metropolitan subway, or bus systems are decried as failing programs that should survive on no public funds. During the last forty years, the country has experienced an explosion in auto use, especially during the years from

1960 to 1980, when the United States embarked on the program of highway construction called the Interstate system. The federal government offered states ninety cents for each ten cents they contributed toward the construction of a national highway network, and the result was an increase in paved road mileage from 1.2 to 2.4 million miles from 1960 to 1998. Total lane miles grew even more dramatically, especially on urban freeways. During the same period, driving grew by over 300 percent, or an average of 9.6 percent per year.

These highway systems, coupled with a change in zoning codes and street design fostered by federal agencies dramatically reconfigured the nation's urban landscape, making it harder to get around by train, bus, or foot. The new highways penetrated the cities, destroying homes and disconnecting neighborhoods from one another. The early increase in vehicle use was seen as a function of the growth in auto ownership, driver licensing, and population; but it is clear that the increase and improvement of highways played a large part as well.

In recent years, it has become clear that suburbanization has contributed to the growth of driving. As people spread out into far-flung suburbs, two things happened: commuters must travel longer distances to work, school, and shopping, and each of those activities becomes more spread out. Thus, the average household stayed roughly the same size from 1983 to 1990, as measured by the Nationwide Personal Transportation Survey, but its auto travel grew by about 12,000 miles per year. Over the same period, the length of the average work trip grew from 8.5 miles to 11.6 miles, as sprawl has moved home and work farther apart. Catherine Ross and Anne Dunning examined land use factors in the National Personal Transportation Study and found that both auto ownership and miles driven per adult increases significantly as population density declines. So we have a classic case of a system out of control, where the provision of physical infrastructure to accommodate travel has stimulated consumption of both vehicles and travel, which has in turn stimulated sprawl, which in turn causes a demand for more highway infrastructure.

In recent years, the growth in travel has been accompanied by a decrease in vehicle efficiency and an increase in harmful emissions. A two-decade trend toward better fuel economy has been reversed and average miles per gallon of the U.S. auto fleet has declined in the past five years as consumers responded to heavy advertising and began to purchase light trucks and sport utility vehicles. These vehicles now comprise more than half of new sales each year. Dramatic improvements in tailpipe emissions of ground-level pollutants brought on by clean-air regulations are being offset by increases in driving. As to gasoline consumption and carbon emissions (which are the main problem for global climate change), we are seeing dramatic declines in efficiency. But the Corporate Average Fuel Economy (CAFÉ) standards haven't been changed since the Carter Administration, and indeed Congress and more recently the Bush Administration have blocked any improvement of these standards, continuing to exempt light trucks and sport utility vehicles from efficiency requirements. The car companies have exploited this loophole.

The impacts of so many Americans driving more and bigger vehicles are profound. More than one-third of U.S. carbon dioxide emissions and 40 percent of nitrous oxide emissions come from the transportation sector. These emissions have contributed to a host of health problems. Recent reports by scientists from the Centers for Disease Control link sprawl and driving to increases in asthma mortality and to a precipitate decline in physical activity, which is in turn a factor in growing rates of heart disease and childhood obesity.

Automobile by-products including brake and tire particulates, air toxins and pollutants, and road chemicals run off into groundwater and are increasingly acknowledged as a major source of both ground and surface water pollution. At the global level, Americans consume more than one-third of the world's gasoline. For every gallon burned, approximately nineteen pounds of carbon dioxide are emitted into the atmosphere, contributing unintentionally but dramatically to melting ice caps and a host of threatening problems tied to global warming.

By some estimates, one-third of our cities are devoted to the automobile in the form of streets and parking lots. Over 41,000

Americans die each year on highways, and Americans are more likely to be killed as pedestrians than they are to be killed by a stranger with a handgun. In fact, autos are the leading killer of the nation's teens. A recent National Research Council report estimates the annual costs of death and injury from automobiles and highways at $182 billion, an annual amount equal to the total cost of building the Interstate system.

OUR APPETITE FOR CONSUMING LAND— A BY-PRODUCT OF AUTO DEPENDENCY?

A recent survey of planning experts and historians found that the most significant planning action of the past century was the construction of the Interstate Highway System, for it made possible the suburbanization of America. In many ways, suburbanization has had a positive influence on American life. Prior to World War II, many urban dwellers lived in overcrowded conditions, with inadequate ventilation, poor emergency access, little heating, and substandard plumbing. The postwar housing boom solved those problems.

But suburbanization also unintentionally caused the rapid loss of farmland and open space, destruction of vital habitat for hundreds of plant and animal species, increased air pollution and climate change, and weakened the fabric of our communities and neighborhoods. In part, this is because the availability of the automobile and the creation of metropolitan highway networks has made it possible for people to live farther and farther away from work, school, and shopping. In part, suburbanization is due to the exploitation of cheaper land on the urban fringe made accessible by the highway network, and so families choose cheaper housing farther out. Zoning and traffic codes from the 1920s, which legislate against mixed land uses and force rigid separation between housing, work, and shopping activities, have made the problem worse.

THE IMPACT ON HOUSEHOLD BUDGETS

The consumption of transportation has a major impact on household budgets for all Americans. The American Automobile

Association estimated the annual cost of owning and operating an automobile at $7,363 in 1999. About 75 percent of that cost is fixed costs, such as car payments and insurance, and this means that there is little financial incentive for drivers to drive less once they make the investment in a car. Nationally, transportation expenditures account for 17.5 percent of the average household's budget, according to an analysis of Bureau of Labor Statistics data by the Surface Transportation Policy Project and the Center for Neighborhood Technology. The proportion of household expenditures that is devoted to transportation has grown as our use of the automobile has grown, from under one dollar out of ten in 1935 to one dollar out of seven in 1960, to almost one dollar out of five from 1972 through today.

The transportation burden borne by American households falls most heavily upon the poor and lower middle class, as the less a family makes, the more of its budget goes to transportation. The poorest quintile of American households spend 36 percent of their budget on transportation, while the richest fifth spend only 14 percent. This means that the poorer a family is, the less money it has available for other expenses such as housing, medical care, or savings. In fact, transportation takes up the second largest percentage of the household budget, ahead of food, education, medical care, and clothing, only behind expenses for housing.

The cost of transportation varies widely from region to region, and within metropolitan areas (see table). Scott Bernstein and Ryan Mooney of the Center for Neighborhood Technology recently analyzed data from the Consumer Expenditure Survey from 1998 to 1999 and revealed that transportation costs can vary from 14 percent of a household's total expenditures in the New York Metropolitan area to as much as 22 percent in Houston.

These differences between regions are due to land use differences. (The difference between New York and Houston would have been greater had not the New York metropolitan area as defined by the Bureau of Labor Statistics included the suburbs of New Jersey.) Research at the metropolitan level done by John Holtzclaw and others shows that sprawl accounts for the difference in transportation costs by forcing an overreliance on the automobile and by necessitating longer trips. This study, which analyzed odometer

COMBINED EXPENDITURES FOR HOUSING AND TRANSPORTATION FOR SELECTED METROPOLITAN AREAS, RANKED BY TOTAL DOLLAR AMOUNT SPENT

RANK	METRO AREA	HOUSING + TRANSPORTATION EXPENDITURES	PERCENT SPENT ON HOUSING + TRANSPORTATION
1	ANCHORAGE, AK	$26,650	52.3%
2	SAN FRANCISCO-OAKLAND-SAN JOSE, CA	$26,245	51.4%
3	SAN DIEGO, CA	$24,820	56.1%
4	WASHINGTON, DC-MD-VA-WV	$24,552	49.3%
5	MINNEAPOLIS-ST.PAUL, MN-WI	$23,820	49.2%
6	DENVER-BOULDER-GREELEY, CO	$23,648	53.2%
7	LOS ANGELES-RIVERSIDE-ORANGE CO, CA	$23,485	54.3%
8	NEW YORK-NORTHERN NJ-LONG ISLAND, NY-NJ-CT	$23,152	52.6%
9	PHILADELPHIA-WILMINGTON-ATLANTIC CITY, PA-NJ-DE-MD	$22,679	56.5%
10	HOUSTON-GALVESTON-BRAZORIA, TX	$22,678	52.7%
11	SEATTLE-TACOMA, WA	$22,549	51.2%
12	PORTLAND-SALEM, OR-WA	$21,964	50.7%
13	ATLANTA, GA	$21,797	56.3%
14	MIAMI-FORT LAUDERDALE, FL	$21,388	56.1%
15	PHOENIX-MESA, AZ	$21,105	52.5%
16	DALLAS-FORT WORTH, TX	$21,063	47.6%
17	BOSTON-WORCESTER-LAWRENCE, MA-NH-ME-CT	$20,923	53.3%
18	MILWAUKEE-RACINE, WI	$20,418	55.4%
19	DETROIT-ANN ARBOR, MI	$20,180	52.9%
20	HONOLULU, HI	$20,026	47.2%
21	CHICAGO-GARY-KENOSHA, IL-IN-WI	$19,965	52.3%
22	BALTIMORE, MD	$19,836	49.9%
23	KANSAS CITY, MO-KS	$19,210	51.9%
24	CLEVELAND-AKRON, OH	$19,077	50.6%
25	CINCINNATI-HAMILTON, OH-KY-IN	$18,929	50.6%
26	ST. LOUIS, MO-IL	$18,318	50.7%
27	TAMPA-ST. PETERSBURG-CLEARWATER, FL	$17,030	51.7%
28	PITTSBURGH, PA	$15,787	45.7%

SOURCE: CENTER FOR NEIGHBORHOOD TECHNOLOGY, CONSUMER EXPENDITURE SURVEY, 2002

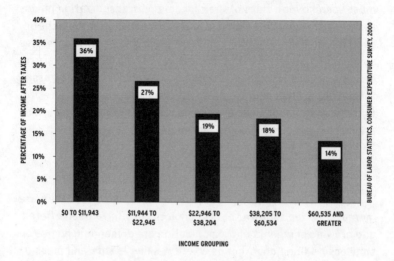

readings collected as part of air-quality inspection and maintenance programs, found that residents of denser, transit-rich neighborhoods drove far less and spent far less on transportation than
people who lived in more sprawling suburban locations characterized by single-use zoning.

THE IMPACT ON WEALTH CREATION

The growing proportion of consumer expenditures that is devoted to transportation inhibits families from devoting their income to saving or investing, and indeed may be part of the reason
why so many families have to send two people to work. For the fact
is that spending on transportation by poor families, unlike spending
on home ownership or investing in education, has a very poor return on investment because autos, unlike houses, are depreciating
assets. Ten thousand dollars invested in a car declines to a value
of about four thousand dollars in ten years' time, while investment
in home ownership builds equity and often appreciates. Similarly,

investment in college education for one's children increases their earning power over their lifetime. The fact that the poorest families must spend over a third of their income on transportation means that they are least able to invest in activities that offer them the opportunity to build wealth. It is indeed ironic that many progressive social scientists believe that the best way to help former welfare recipients secure jobs is to give them automobile purchase assistance, thereby trapping them into the poverty cycle even more profoundly, as the poor typically end up with less reliable cars, which are more expensive to operate and maintain.

THE WAY OUT–ENCOURAGING TRENDS

That's the bad news. The good news is that Americans are increasingly fed up with the bad choices they have been offered, and growing segments of the population are demanding homes on smaller lots in neighborhoods where walking is safe and pleasurable and amenities are within walking distance. This shift in preference is being seen by developers, builders, and elected officials, and they are beginning to respond by building mixed-use neighborhoods, reviving urban neighborhoods and walkable suburbs, and building dozens of new transit systems.

GROWTH IN DRIVING FLATTENS: TRANSIT USE IS ON THE RISE

One of the most hopeful signs is that over the past two or three years, the upward slope of the driving trend has begun to flatten, even as population has continued to rise. The rate of growth in driving has dropped from 3 percent per year in 1998 to a slight decline in 2000. At the same time, public transit ridership is increasing. Transit's heightened popularity comes against a much smaller base, partially due to the fact that public transportation is only available to about half the population, and available and convenient for a much smaller percentage. Transit use grew 11 percent from 1998 to 2000, according to ridership statistics collected by the American Public Transit Association. Much of this growth was concentrated in cities where lots of transit service was provided like New York City and Chicago, but 2000 saw significant ridership increases in cities in nontraditional places, too. Bus ridership grew in Oklahoma City by almost 8 percent, in El Paso by over 13 percent, and Spokane by 7 percent from 1999 to 2000. Some of this growth in transit use is due to the fact that transit agencies are learning to make riding more convenient for the passenger by offering multi-ride smart cards, but much of it is due to the fact that transit is becoming more available as the roads become more congested. New rail systems are proliferating all over the country, with new rail lines opening in the past decade in places far from the transit rich East Coast such as St. Louis, San Jose, Denver, Los Angeles, Dallas, Portland, Salt Lake City, Sacramento, and San Diego. All of these cities have met or exceeded patronage forecasts, and all of them have plans underway to expand their systems.

New transit systems are in the planning or construction stages in a host of other metropolitan areas around the country including Houston, Seattle, Minneapolis–St. Paul, Phoenix, and Tampa. Most of these new systems are commuter rail or light rail systems, although a growing number are busways or rapid bus systems. Both types offer the rider the option of a vehicle that has time advantages over the private motor vehicle, either because they don't have to share the roadway with cars, or they are given preferential treatment at traffic signals.

LIVING DOWNTOWN–A GROWING TREND

One of the other recent trends that may indicate a move away from the growth in consumption of land and auto use is the growing number of people who are choosing to live in downtown areas. A recent study by Rebecca Sohmer and Robert Lang for the Fannie Mae Foundation and the Brookings Institution found that sixteen out of twenty-four downtowns surveyed had grown in population from 1990 to 2000. While this trend is still small in comparison to the number of households choosing to locate on the periphery of metropolitan regions, it may be an indicator of people's willingness to live in denser areas. Indeed, downtown areas are increasingly seen as places to live, not just as places to work and shop. Sohmer and Lang argued that the desirability of downtowns is due to their proximity to work, mass transit, and amenities, and that this proximity augured well for a continued growth in downtown populations, and for perhaps a spillover into adjacent urban neighborhoods. One of the downtowns to have experienced the most growth in Sohmer and Lang's study was Chicago, which grew in both downtown population and density by almost 50 percent in the decade from 1990 to 2000.

PRIVATE MARKETS ARE RESPONDING

This growing trend back into more compact, walkable, mixed-use communities is being driven largely by consumer preferences in the marketplace, and not by government action. Increasingly, real estate developers are seeing an untapped market in providing new housing types. Heralded as the "new urbanism," this visionary development has attracted architects and builders, because it is beginning to offer an attractive alternative to urban sprawl. The new urbanism also provides a comprehensive template for development that appears to attract a premium in the marketplace, according to the Urban Land Institute. Hundreds of new urbanist developments are open or underway across the country.

As a consequence, real estate forecasters and investment experts are advising their clients to invest in mixed-use communities, and companies are showing a preference for these kinds of developments. A recent study by Jones Lang LaSalle of so-called "new economy" companies found that access to mass transit was a very important factor in location selection for 77 percent of companies surveyed. The annual Emerging Trends in Real Estate report, which rates all types of real estate investment in differing metropolitan areas, advises investors to select locations characterized as "24-hour cities," with mixed use development and access to transit. Investors are cautioned to avoid investing in projects in suburban locations without access to transit, as growing congestion makes these risky investments.

The private marketplace has also seen a growing trend toward transit-oriented development, which is located and designed to take advantage of proximity to mass transit. Transit-oriented developments are located within easy walking distance of a transit facility; contain a mix of uses including housing, services, and retail; and are designed to make driving unnecessary for many trips. Often spurred by local government investment in transit, these kinds of developments are finding easy acceptance among consumers. One developer in the Portland, Oregon, area put it this way: "We had forty-six sales in the first four months, which is the highest absorption we've ever achieved for a new product in the Portland market."

Even those living in dense urban neighborhoods have an occasional need for an automobile. In response, a variety of schemes for car sharing have emerged in Europe and the United States. Typically, a variety of cars are purchased and distributed throughout neighborhood locations, and car-sharing customers or members are able to access them within an easy walk or bus ride of their home, in contrast to auto rentals, which are concentrated at airports and in downtown locations. Car sharing, which is being offered under a number of different business models, including hourly rental and member cooperatives, is moving rapidly into the American market. A recent survey found that car-sharing programs were being introduced in about a dozen American cities, including Chicago, Seattle, and San Francisco, and that car rentals were also considering adding the hourly rental feature to their product lines.

Some lending institutions are also changing loan criteria to reflect changing consumer preferences and needs. The Location Efficient Mortgage (SM), a product of Fannie Mae and a consortium of groups called the Institute of Location Efficiency, allows prospective home buyers in denser transit-rich neighborhoods to use their transportation savings to help them afford a home in these neighborhoods. The program, which has been introduced in Chicago, Seattle, and San Francisco, is expanding to Atlanta, Portland, and Philadelphia, and Fannie Mae has announced plans to introduce a less comprehensive product with smaller savings in Minneapolis-

St. Paul and Baltimore. In essence, financial institutions are now sending a message—if you save money by driving less, we'll take that into account and offer you more funds to purchase a home.

CHANGING DEMOGRAPHICS MEAN CHANGING DEMAND

These products are responding to changes in consumer preferences. Cities, once stigmatized as crime-ridden repositories of the poor, are now being seen as vital, resource-rich places, in part because increased density creates the opportunity for a more diverse mix of amenities than is available in one-dimensional suburban locations. A larger trend lies just underneath this change in attitude, though. The demographics of the country are gradually shifting, and these shifts portend a fundamental change in the demand for housing and community. There are four interrelated demographic trends underway, which have been dramatically illuminated in the 2000 Census results. Each of them holds the possibility of helping us move from suburban sprawl and traffic nightmares to reinvigorated urban centers with a high quality of life.

Immigration

The most notable finding of the 2000 Census was the unequivocal diversity added to our nation as a result of immigration from other countries, principally Hispanic and Asian households. Historically, most immigrants and most minorities live in cities, and while there is a significant trend toward minority migration to the suburb, demographer William Frey projects that most immigrants will continue to be concentrated in more dense urban locations.

This urban concentration, along with the lower income levels of most immigrant households, has historically meant that these households own fewer automobiles and drive less. According to Catherine Ross and Anne Dunning's analysis of the 1995 National Personal Transportation Survey, African Americans, Asians, and Hispanics are all more likely to use public transit or walk than

Whites. For immigrants, this may be due not only to income and poverty level, but also to cultural factors, including the fact that they have lived in places where transit use was the norm rather than the exception. As immigrants assimilate into the population, therefore, we can expect to see higher levels of driving as incomes rise, but also a continued willingness to use public transit, particularly if its availability, quality, and convenience continue to increase.

"Empty Nesters" and "Echo Boomers"

The second demographic trend is the aging of the baby boom generation, and its passage from the child-rearing stage of the life cycle to the "empty-nest" phase. Families that once demanded the single-family home on a quarter acre parcel in a suburban location are now finding both the home and the location to be unsuited for a new stage of life. Evidence suggests that baby boomers have fueled much of the downtown population growth over the past decade, as they seek smaller homes in locations with a greater mix of amenities.

Marketing experts and demographers alike have trumpeted the echo boomers' (aged twenty-four to thirty-four) preferences for exciting, dense, urban locations. Indeed, the much-publicized growth of new-economy cities like San Francisco and Austin was ascribed to their attractiveness to highly skilled young workers. A recent study found that 57 percent of this generation preferred small-lot housing and that 53 percent felt that an easy walk to stores was an extremely important determinant in housing and neighborhood choice.

Nonfamily Households

The 2000 census found that nonfamily households comprise 31.9 percent of all American households, more than married couples with children at home, a group that now comprises only 29.5 percent of households. Ross and Dunning found that single adults with no children and households of two or more adults with no children were most likely to live in urban locations. These less conventional households are another force for positive change.

THE SHIFT IN HOUSING AND NEIGHBORHOOD PREFERENCES

These demographic trends add up to a growing market for smaller homes, town homes, and homes on smaller lots, and a desire for more vibrant, walkable neighborhoods. In a recent study released by the Congress for the New Urbanism, Dowell Myers at the University of Southern California estimated that between 30 and 55 percent of the demand for new housing would be for residences in dense, walkable neighborhoods. He also found that almost 25 percent of the aging baby boomer demand was for town homes in the city.

MEETING THE DEMAND: THE ROLE OF PUBLIC POLICY

If we hope to contain and even reduce the number of miles Americans drive alone each year, we must first acknowledge that the auto will remain the single dominant mode of travel for the foreseeable future. The automobile is convenient, reliable, and cheap. At present, most Americans do not have a choice not to drive for most of their trips. Walking or bicycling is difficult if the streets lack sidewalks or if shopping, schools, and libraries are inaccessible. Public transit is often not within reach, and when it is, it often is more expensive than driving and less frequent than is convenient. The cities and suburbs where most of us live are structured around the automobile—our zoning codes, financing systems, and tax laws encourage developers to build single family homes on large lots that are physically separated from daily activities. Originally, these laws were meant to protect us from smelly factories, but now they separate us from a loaf of bread, the hardware store, and the elementary school. A multifaceted approach to this problem is essential—an approach that gives Americans a choice not to drive rather than an approach that seeks to punish us for driving. At the same time we need to build consciousness about the need for improved efficiency and Detroit should be required to offer more fuel-efficient vehicles at all levels of the automotive fleet.

In large part, the effort to reduce driving is also an effort to restore communities to places we'd like to live and share in, rather than transient places to build equity. Creating livable and walkable communities where transit is an option can increase our access to opportunities and amenities, serving a variety of ends. A shift to these kinds of communities can help families build wealth, as less income is devoted to depreciating assets like cars, and more is available for home ownership or saving for education. At the same time, these diverse, mixed-use neighborhoods are more interesting, and families will find that the less time they spend commuting or carpooling children around, the more time they have for their children.

GIVING AMERICANS THE CHOICE NOT TO DRIVE

What will it take to make this vision a reality: to reduce driving and improve efficiency in a manner that increases opportunity and enhances quality of life? A truly sustainable system will require a variety of positive changes. There are a host of policy reforms that could move us in the right direction. These include home mortgages reflecting reduced expenses from living in a dense, transit-rich neighborhood. Or tax policy that provides as favorable treatment for people who bike or take transit to work as it currently does for those who get free parking from their company. A collaborative national effort by the American Planning Association has just been completed to develop new zoning and planning standards that favor mixed-use development that is less auto-dependent to replace the old standards that have brought us the suburb. Government would do its part by locating government facilities and service in transit-oriented communities, by linking affordable housing subsidies to such communities, and by encouraging its workers to telecommute. Over time, these changes would result in the development of communities that are accessible by more than automobiles, where transit really works, and where neighborhood businesses reduce the need for much travel.

At the same time, we need to build the infrastructure that makes all this possible. The 1991 Intermodal Surface Transportation Efficiency Act and its 1998 successor TEA-21 have already provided cities, suburbs, and states with the opportunity to invest in travel options other than highway development. Under this federal law, local governments can opt to allocate federal transportation funds to any number of transportation options, from bike paths to hybrid electric buses. Increasingly, they are doing so, as they realize that to cure congestion in the suburbs by widening roads is like trying to cure obesity by loosening one's belt. In the near future, affordable, frequent, reliable bus or rail transit will be a short walk away in the central parts of our major cities and "smart" shuttles and vans are a phone call and a short wait away in our suburbs and smaller cities and towns. Walking should be an option, with pedestrian routes no longer being considered a hindrance to car travel. A network of bike trails, paths, and lanes would be almost ubiquitous, using available rights of way in streets and abandoned rail corridors. The federal transportation law is up for renewal in 2003, and this presents an opportunity to do more to encourage states and localities to invest in transportation facilities that encourage community livability, make walking and biking safe and pleasant as well as meet the overwhelming demand for new transit systems.

Advances in technology and communication, coupled with changes in the workplace and the economy, could help, too. I foresee a steady rise in the number of knowledge workers and a steady increase in the percentage of these workers who "telecommute" to their offices from home using computer, fax, phone, and modem. Similarly, experts believe that many shopping and personal business trips—banking, paying bills, etc.—will be obviated by the use of home-based shopping by phone, television, or computer. If this is coupled with more walking opportunities for daily shopping and interaction, the result could be quite powerful. Of course, this is by no means certain, as studies have shown that telecommuting on its own results in work trips being replaced by other trips. But all of these options are intended to work together, providing a major shift in the convenience and accessibility of our communities. The impact would not be seen overnight, but would be gradual, over a few decades, just as the shift to suburban auto dominance has been gradual. Ultimately, however, the impact would be profound, both in reducing consumption of fossil fuels and in meeting other goals of a sustainable society, such as social equity, job creation, and enhanced quality of life.

To accomplish this "retrofitting of cities, suburbs, and towns," we need a deliberate and gradual removal of government regulations and subsidies that favor decentralization and single-use, auto-dependent development. In their place we would put a set of policies that encourage resource-efficient, mixed-use communities. This is not an attempt to outlaw single-family suburban developments, indeed the suburb will continue to be the choice of many. But people should have a realistic, affordable choice of other kinds of communities. The market distortions that prevent such communities from developing should be eliminated. Furthermore, suburban communities can be reconfigured to be more efficient with mixed-use community centers and improved pedestrian and transit accessibility. After all, the work trip is a minority of all trips. Improving accessibility of shopping, schools, and services can help to reduce nonwork auto travel. Along with more efficient suburban growth patterns, increasing numbers of people may prefer a more urban lifestyle. Policies should ensure that well-developed, livable communities are available in urban areas. This may entail both redevelopment and renovation of existing neighborhoods and new development in former industrial areas and around existing or new transit stations and downtown living.

A similar opportunity exists for the redevelopment of older, inner-ring suburbs. Tremendous private and public capital is at risk of being stranded in the older shopping centers, strip malls, offices, and housing subdivisions of these aging suburbs. Suburban jurisdictions and investors alike need a way to preserve the value of these investments. High-access, livable places may be the solution. This goal would be accomplished through a variety of strategies—

acting in an integrated fashion with each other to provide a choice of communities and in concert with transit and other supply options to provide an alternative to solo driving. To this end, we need:

- Continued research and public education on the quality of life impacts of auto- and fossil fuel-dependent lifestyles. We need to document and publicize hours behind the wheel, the impacts of sedentary and obese lifestyles our children are being taught, pedestrian death tolls, the rise of road rage and aggressive driving, the increased percentage of our disposable income being expended on transportation, and the health impacts of the automobile.

- Enactment of strengthened Corporate Average Fuel Economy Standards. The CAFÉ Standard should rise to forty miles per gallon over the next several years, and the exemption for sport utilities and light trucks must be eliminated. These vehicles could be phased in at some achievable yet tougher standard.

- A government led *public/private mixed use-financing initiative* to identify and end barriers to financing of higher density, mixed use development.

- Expansion of the program for *location efficient mortgages.* Fannie Mae has announced market tests of this program in Chicago, Los Angeles, and Seattle.

- Extension of the regional and statewide planning structure developed under ISTEA and TEA-21 to other programs that influence and guide growth and development—HUD's new block grants, HHS service grants, etc. All should be linked to a *regional structure for metropolitan planning* so that housing, business development, and service delivery can be regionally designed and delivered as part of regional growth strategies.

When I was growing up, getting a driver's license and access to the family car meant freedom and getting out of the house. Some part of me still carries that association, but the larger part associates driving with traffic jams, nonproductive time, and expense. More and more, we have nowhere to go in our cars. If we surrender our towns, countryside, and cities to the car, we will also be surrendering many other values that we hold dear: neighborhood life, a sense of history and place, a feeling of belonging somewhere.

IN PRAISE OF HOMETOWNS

Mary Pipher

Communities exist for the health and enjoyment of those who live in them, not for the convenience of those who drive through them, fly over them or exploit their real estate for profit.

—*Theodore Roszak*

NINE MILE PRAIRIE

Early evening, I drive with my friend Pam past crowded malls, cinemas, and fast-food joints. Pam has been my friend for thirty years. Her husband and mine have been best friends since kindergarten and Pam sings with my husband's bands. We've had babies together.

We drive to where there is plenty of quiet space and park in an empty lot. We step out into the grainy thick air to hear the trill of larks and the cries of bats trolling for mosquitoes. I ask myself, as I always do at this place, "Where is everybody? Don't people realize what we have right under our noses?" On the other hand, I am glad everybody is not here. For a few hours, Pam and I have the luxury of a world without humans.

Tonight the tall grasses are backlit by the sun and flame against the green cottonwoods. It's the time of day that wise traders sell horses. With their coats burnished by the sun, they're irresistible. Rose hips sparkle like Christmas decorations in the bluestem and gamma grass. Wild plums, a frosty purple color, hang thick over fences. We've picked them many times to make Chinese plum sauce. Rose mallow and goldenrod are blooming. Rare prairie orchids should be blooming, but they are hard to find.

Sometimes we walk on paths and sometimes we tromp through ten-foot tall grasses where we lose each other and shout to identify our location. We fall down on purpose and allow red waves of grass to roll over us. Back on the paths, we hear and then see an eastern bluebird. He flies ahead of us, stops, then when we catch up with him, he leads us on. At dusk, the first few fireflies light their lamps, creating the illusion of a little prairie town. To the east, the lights of Lincoln twinkle on. The dome of the capitol is a rosy-gold in the sunset. We watch our town light up, then continue our search for prairie orchids.

Pam and I had busy weeks and, like all humans, we have our problems, but as we walk here, breathing in the aroma of the grasses and flowers, feeling the scratchy seedheads and the soft breeze, seeing the blue-pink sky and hearing the animals at work, these problems fall away. We grow quieter. Our breathing changes. We feel at home.

This night turns out to be a night of surprises. As we walk west toward our favorite ridge, we notice dozens of mating monarchs. The males are flying above the females, holding their wings together and uniting in the air.

We stand and watch. We can almost touch them, so oblivious are they to our presence. One couple tangles Pam's hair and five couples land on the Joe Pyeweed plant by my side. We are awash in monarchs.

We congratulate each other on being here on monarch date night. We follow them west right into a small glen, where we discover some prairie orchids blooming. They are tiny white flowers with petals like flags swirling around a delicate silver pole. We breathe in their fragrance. Prairie aromatherapy.

We hike to the ridge and watch the red sun melt into the ground. High above us to the north a red-tailed hawk circles. The lightning bugs continue to switch on, mostly down in the glen among the cottonwoods and sumac, and beside us in the tall grasses.

I wonder, as I always do, what this land looked like to the Native Americans, what our country was like when it was, to quote Paul Gruchow, "a great green heart." The pioneers talked of birds so thick they blocked out the sun, of wild strawberries so plentiful they stained horses' fetlocks red.

Slowly we walk toward our car, quieter now and calmer, our worries put in perspective, our arousal systems tamped by that greatest of all antianxiety drugs—a sunset.

Nature always speaks if one listens. Tonight she whispered to me, "The world is bigger and older, and more important than you are. It is stronger, more fragile, more complex and beautiful than you can comprehend. And yet, you are part of it. Take care of it and let it take care of you."

Once when I was in New York City, a woman asked me where I was from? When I answered, "Nebraska," she asked rather rudely, "Have you considered moving?"

I mumbled an inane answer, but I have been thinking of a better answer ever since. My answer tonight would be, "No. I am home and home is not a salable piece of real estate."

It's that simple and that complicated. Nebraska for most people is the state they drive through on the way to someplace else. Our state is six hours of Interstate 80. Of course, that's not the way to see any place.

In general, Nebraska is too stark to be pretty, but it's beautiful. It's beauty is its scope, its great table of horizon, its skies that dominate all earthly landscapes, and its great rivers; the Loup, the Dismal, the Niobrara, and the Platte. Nebraska contains the vast and quiet Sandhills, where the population is less then one person per square mile. It shelters cranes, meadowlarks, and mourning doves. With our state, the trick is knowing how to find its beauty. Once when someone said Nebraska wasn't beautiful, my husband responded, "Come back with a better pair of eyes." But Nebraska isn't my home because of its beauty. It is beautiful because it is my home. The curve of its hills and the songs of its cicadas have etched themselves into my mind. The landscape of my childhood is the Nebraska horizon. When I am away from here for more then a few days, I yearn for the sights and smells of Nebraska. Wherever I travel, I look for the geese overhead, the empty spaces, the cottonwood and Russian olives that remind me of home.

As Eudora Welty said, "As soon as a man stopped wandering and stood still and looked around him, he found a god in that place." That's how I feel. Anyplace can be home, can be beautiful, if you stop and claim it, if you take the trouble to discover what is available to love.

ONE BIG TOWN

Soon the question where do you come from will be as antiquated as what regiment do you belong to?

—*Pico Iyer*

Demographic clusters have replaced national identity as the great definers. People in those clusters share the same activities, opinions, and tastes whether they live in London, Milan, Hong Kong, or Lincoln. Everywhere is becoming everywhere else. Globalization, war, environmental catastrophes, and mass migration have led to an upending of cultures that affects all of us. Our world is often referred to as a global village, but it could perhaps be more accurately described as a global strip mall. It's tawdry, impersonal, and dull. Globalization means we all live in one ugly company town. Many of us are trying to find a way back to a place called home.

In the past century, the Midwest where I live has undergone enormous demographic changes. All over the prairie, the lights have gone out as farmers have moved to the suburbs and little towns have dried up like tumbleweeds. Downtown cafes have closed and the locals now drink coffee at the Arby's on the highway. As we travel the interstates, which Paul Gruchow called "tunnels without walls," we see the same stores, cafes, and hotels everywhere.

Bill McKibben defined a working community as one in which it would be difficult for outsiders to fit in. That's because the information in the community would be specific, related to that time and place and grounded in the history of it inhabitants. Songwriter Greg Brown said, "Your home town is where you know what the deal is. You may not like it, but you understand it. You know the rules and who is breaking them."

When I think of a working community, I think of my father's Ozark town. Cousins lived near each other and everyone knew everyone. Outsiders had a tough time getting information about locals because the only outsiders were salespeople or IRS and FBI agents looking for moonshine stills. On the other hand, sixty years after my father left the Ozarks, I can still go there, explain who my family was, and extract special privileges—a campground on private property and advice on where to fish and pick berries.

Strip malls and sprawl have taken their toll on sense of place. Similarly, our host of new inventions, gadgets, and technological tools have undermined community. While many of these inventions have improved our lives in some ways, they have also eroded the fabric of family and community. For example, air-conditioning has changed neighborhoods. Adults no longer sit on their front porches to cool down in the evenings. Streets have become more dangerous without the supervision of neighbors. Automatic dishwashers have saved time for many women, but they have also eliminated time after dinner when family members worked together and talked.

New tools have sped up the pace of our lives. When people communicate by E-mail and fax, the nature of human interaction changes. Laptops and cell phones allow us to work all the time, and many people do. All the technology of our times has its good uses, and any one invention probably wouldn't do that much damage; the problem is the whole pile. The cumulative effect of all this equipment has changed our daily family life. Quantity has replaced quality and the integrity of our lives has been altered.

Television, among all the inventions of the past fifty years, has had the most significant effects on community. Information and entertainment come from boxes, not neighbors. People spend more time watching music videos, but less time making music with each other. People in small towns now watch international cable networks instead of driving to their neighbor's house for cards. People in urban settings watch soaps instead of visiting local art galleries and museums. When company comes, the kids are sent to the TV room with videos. Television is on during meals and kids study to television or radio.

Some of the first voices children hear are from the television and the first street they know is Sesame Street. Children learn different messages from these boxes than they would learn from loving adults. Everything from their social skills to their moral development to their coping strategies is different. TV-watching children have short attention spans and long want lists, at the same time that they have poor impulse control and fewer real skills. Not surprisingly, we have an epidemic of childhood depression.

Television and electronic media have created communities with entirely different rules and structures than the ones of the past. Families gather around the glow of the TV as the Lakota once gathered around the glow of a fire on the Great Plains. But our TVs do not keep us warm, safe, and together. Rapidly, our technology is creating a new kind of human being, one who is plugged into machines instead of relationships, one who lives in a virtual reality rather than a family. And just as families have unraveled, so have communities.

This unraveling of our public life and fracturing of our communities has many costs. Social service agencies are overwhelmed as families can no longer take care of their own. Children are more afraid. They are warned to avoid strangers and, in electronic villages, everyone is a stranger. Old people are isolated from the young. Children yearn for more lap time and teens develop poisonous peer cultures. Families don't have the social resources they need to raise healthy, wholesome children. Many people are lonely. School-bond issues don't pass because the older generation doesn't know children and has only a limited investment in their rearing. Land and water are not protected because we are educated to accumulate private wealth and permit public squalor.

Margaret Mead defined the ideal culture as one in which there was a place for every human gift. No better definition of an ideal culture has ever been written. It includes both respect for the individual and belief in the ability of communities to foster growth in their members. It is hard to realize the gifts of people whom we do not know. It is impossible to develop our own gifts without a web of human relationships. It is also harder to be kind. Because we don't know the people with whom we are interacting, we can't inquire about their problems or empathize with their troubles. We don't notice that they look stressed or tired. We can't congratulate them on their children's victories. If we see social interactions as the web that holds our lives in place, that web is torn and tattered by the effects of our technology.

HOME IS OTHER PEOPLE

Joy Harjo wrote, "As long as there is respect and acknowledgment of connections, things continue working. When that stops we all die."

Home doesn't have to be where you were born or grew up. It doesn't have to be a small town; it can be a suburb, a city, or a remote corner of the country. But it does have to be a real place that you have committed to over time. It has to be a place where you have friends and know the names of many people you meet. You know who is kind and honest, who lies, betrays, or fools around. Home is where people care if you have a speeding ticket or a fever. It's where people ask about your grandbaby and your day lilies and know your favorite kind of pie. It's where when you sit down to talk you don't have to discuss Tom Hanks or Benecio del Toro. You have real people in common.

In true communities, conversation is often, "Did you know he was John's son-in-law?" Or, "Is she related to the Carsten family?" "Did you know they bought the old Walinshinsky place?" Or "Did you know Jane and Betsy were schoolmates?" There is pleasure in just acknowledging each other, in nodding on the street and chatting in the cafes and grocery stores. To move away from a true home is to move away from life. I don't think we begin to acknowledge and understand how much we have lost.

Yet, the great irony is that people move away from home all the time. Every year our young adults head for the mountains, the Sunbelt, or the coasts. Many leave of necessity, especially from farming communities where there is no longer much work. But many leave because the grass looks greener in San Francisco, Phoenix, or Missoula. Likewise older citizens are encouraged by advertisers to move to retirement areas in more beautiful places far from home. When they get there, they are often alone. As Greg Brown said, "You can't drink a cup of coffee with the landscape."

I know a family who moved back to New York City after September 11. The wife told me, "That is where our people are. We need to be with them." A gay couple I know have inspired many of their friends to move to the outskirts of Omaha. They come there not for Big Red football or the steaks, but for the camaraderie and mutual support of their long-term buddies. I know a couple whose son and daughter-in-law lived in California. When the daughter-in-law became pregnant, they wrote the couple a "letter of invitation." They didn't want to pressure their son or his wife to return to Nebraska but they wanted them to know how welcome they would be. They wanted to remind the young family about what was good in our state. In the letter, they mentioned the safety, the ease of travel, the low cost of living, the wide-open spaces, and the good educational system. They wrote about the love they would give a grandchild and the ways the family could help each other if they shared a community. Much to the joy of the parents, the children decided to come home.

Communities are real places, chosen as objects of love, with particular landscapes, sounds, and smells and particular people who live there. Communities are about accountability, about what we can and should do for each other. People who live together have something that is fragile and easily destroyed by a lack of civility. Behavior matters. Protocol is important. Relationships are not disposable. People are careful what they say in real communities because they will live with their words until they die of old age.

Connections have a way of making us morally accountable. At a most basic level we behave better with people and places we will see again and again. Some of the worst behaviors in America occur in airports and on interstates, places where we move among strangers. Over the Internet, people can be deleted the second they become annoying or tiresome. Names aren't necessarily even real names. One never need see or talk to anyone again.

Responsibility is directly related to scale. The smaller the group, the stronger each member's sense of duty. Morality is learned by children from real people who are with them every day. They learn that their actions affect other people and they learn that their own lives will go better if they behave well. It is a simple thing, to be in a place where good behavior is rewarded and bad is punished. In that sense all morality, like all politics, is local. The further we are from home, from our people, the less likely we are to see a strong connection between our own behavior and its consequences. There is no accountability in a global village except a ledger sheet. And money is not morality.

Strong communities also treasure and maintain the special names, stories, and history that define particular places. Bill Holm wrote, "We stand on the shoulders of our ancestors no matter how many machines we invent. Only our memory and our metaphors carry us forward, not our money, our gadgets, or our opinions." Names, stories, and history are intertwined. We cannot love what we cannot name. One of the best ways to instill community is to teach names—of local people, birds, plants, rivers, and sacred sites. Communities are much enriched by local history books. Older people who are, in a sense, living history books, greatly benefit a community by telling stories to its children. Yellow Springs, Ohio, had an environmental mentor program in which an older person was paired with a young person. The pair walked around town and talked about the town's stories and how the place used to be.

We owe a great deal to people who came before us and to what Paul Tillich called "the structure of grace in history." Communities need ways of sharing stories. This is one of the most primal experiences of humans, to be together telling stories of the day. To be a member of a community is to have a voice and a face in that community. One must be a part of the legends, the colorful characters and the heroes who help define a place.

Story sharing and face-to-face interactions are greatly facilitated by design. Community occurs where there are public spaces—sidewalks, bike trails, parks, outdoor markets, and festivals. Ray Oldenberg wrote in *The Great Good Place,* there are three essential places—where we live, where we work, and where we gather together for conviviality. Those communal places are needed now more than ever.

Diversity in community is as healthy as diversity in any ecosystem. Without diversity in age, ethnicity, and ideas, we don't have communities; we have lifestyle enclaves. Community does not mean "free of conflict." It's inevitable and even healthy to have great differences. Even conflict can lead to closeness. As Dennis Schmitz wrote, "Humans wrestle with each other, and sometimes that wrestling turns into embracing."

A strong community will include people of different ages, ethnic backgrounds, socioeconomic status, and interests. Community, communication, and communion all come from the same word, meaning together and next to. Embedded in the word is the concept of shared place.

RECLAIMING OUR COMMUNITIES

The love of your own country hasn't to do with foreign politics, burning flags or the Maginot Line against immigrants at the border. It has to do with light on a hillside, the fat belly of a local trout, and the smell of new-mown hay.

—*Bill Holm*

On the cusp of a globalized world, at the start of a new century, we can only just imagine a world with no real countries. But the paradox is that the more one travels and has contact with the world, the more one needs a home. The more we live in a global shopping mall, the more important it is to look at the stars and visit with our neighbors. The cure to the cultural colonialism of global shopping malls is loving our home town. "Provincial" and "parochial" have traditionally had negative connotations, but they can also mean the sacredness of one's town. Terry Tempest Williams wrote, "It just may be that the most radical act we can commit is to stay home."

Falling in love with a place is like falling in love with a person; once you are in deep it's best to have a long-term committed relationship, one that requires sacrifice. Commitment implies protecting, nurturing, and defending. Without geography, there is no accountability. Without geography there is nothing to defend.

Many of the things that would change the quality of our neighborhoods are simple, but not easy. Architects could design homes that reach out into the community, homes with front porches or homes built around courtyards so that children can play together in protected places. Developers can save wild areas for children and set aside common space for families to enjoy. Co-housing, which is a way of designing neighborhoods so that communal life is possible, is an important new option. Walking, biking, and good public transportation systems could replace automobiles, one of the greatest destroyers of neighborhoods. The essential step is really the change in attitude, from the worship of privacy to an acknowledgment of our needs for connection, and from this new awareness, architecture will follow.

City apartment buildings can have parties in the lobbies and ways for residents to interact on a regular basis. Urban dwellers can schedule weekly potlucks with friends on the beach or in a park at sunset. Schools can serve a free meal to the school community once a month so everyone knows everyone and their children.

Very simple steps will help you build community. Turn off your machines, walk outside, and talk to the children in your neighborhood. Ask the waiter at the place you buy your morning coffee about his family. Go to a school board meeting, befriend the older people in your condo, coach a ball team, or mentor an immigrant family.

Plant a community garden, learn the names of your neighbors and storekeepers. Buy locally. Know all local flowers. Give schoolchildren trees to plant and encourage them to stay and watch the trees grow tall. Celebrate the first corn of the year and the first urban dogwood blossoms. Sponsor festivals, feasts, and block parties. Go folk dancing, attend bluegrass festivals and enjoy local ethnic celebrations. Sing in a local choir, run for office, start a book club, or form a conversation cafe.

Simple tools help people know each other. For example, when we moved into our neighborhood, the man next door drew us a map of our street. On every house, Ron wrote the names of the parents and what they did, the names and ages of the children, and their phone numbers. He knew which parents were home during the day and which families had pets. This small act made us feel welcome and connected from day one. Later, Ron gave us a list of unusual or expensive items that neighbors were willing to loan. Over the years we shared ladders, microphones, video cameras, snow blowers, pickups, Halloween costumes, and wheelbarrows. These small matters helped a great deal. Once we knew our neighbors' names, we had something to build upon. Sharing goods saved us all money and gave us a sense of connection.

Even the busiest person can do a little. Adults can learn the names of the people on their block or in their apartment building and wave to the children and retired people. They could help kids find lumber to build a fort or help an elderly neighbor into her apartment with her groceries. Adults can talk for a few minutes to the children who deliver newspapers or flyers. Parents of teenagers can meet once a month and talk about how the children are doing. Families can put their lawn chairs in the front yards rather than on fenced-in back patios. Families can organize block parties—potlucks, parades, or star-gazing evenings. Families can carry zucchini bread or garden flowers to each other's houses. They can offer to watch each other's children and, when the children are over, actually play games with them or teach them something. Families could offer to help their neighbors with home improvement projects or write them notes of news, sympathy, or congratulations. Children can set up lemonade stands and all the adults can stop for drinks. It is also good to ask others to help your family. This asking for help also builds community bonds.

Many people protest that they would like to be in a community but they are too busy. They don't have time for other people. What they don't realize is that communities eventually give time to their members. In a community, parents have other adults helping them raise their kids. Neighbors can depend on each other for friendship and support. Fun and companionship are close at hand and do not require elaborate arrangements and lots of money.

Elizabeth Barrett Browning said, "The earth is crammed with heaven." Nine Mile Prairie is one slice of heaven. The coffee-colored Platte another. Every town and city has many places worthy of love and protection. Your slice of heaven may be the Iron Range, or North Beach, or Central Park, or Chesapeake Bay, or Harvard Square. Join with your neighbors to enjoy those places and work to keep them for your great-grandchildren.

TOWARD PROPERTY AS SHARE: OWNERSHIP, COMMUNITY, AND THE ENVIRONMENT

Prasannan Parthasarathi

Although I have lived in the United States for most of my life, I have maintained close ties with India, the land of my birth. This dual existence has made me deeply aware of the vast inequalities in wealth, income, and life chances in the world. In my mind, images of American abundance have long sat uneasily with those of Indian want. As a historian and an economist, I have devoted my scholarly efforts to understanding the origins of poverty in India. As a human being, I have been committed to its eradication.

In recent years, I have become increasingly concerned about two other problems that we face as we enter the twenty-first century. The first is massive environmental destruction. In the last one hundred years, species extinction, deforestation, water and air pollution, and other ecological degradations have proceeded at accelerating rates. We human beings are emitting such massive quantities of poisonous substances into our environment that many parts of our planet may one day become uninhabitable. Alongside this ecological crisis, many Americans feel a heart-aching sense of isolation. We live in a world that is increasingly devoid of deep social connections and there is a widespread desire to recapture a sense of community that—correctly in my opinion—we believe existed in the past.

Poverty, environmental destruction, and loss of community may appear to be unrelated problems, but they have their origins in a common institution: the system of individual private property. The connection between private property and a range of contemporary ills remains largely unexplored because questions of ownership are not the topic of much discussion these days. However, questions of property, including how it is owned and who owns it, must be put back on the political agenda. For the institutions of property not only shape our economic lives, but also our social existence and our relationship to nature.

In an economic order centered on private property, individuals are allowed to enjoy and use their property as they see fit with little regard for the well-being of others. They are encouraged to act selfishly and are highly rewarded when they do so. Rather than seeing relationships with others as essential to human existence and fulfillment, other people become potential obstacles to the taking of profits from one's property. In such a world, it is not surprising that the possibilities for meaningful communities are limited.

In a system of private property, protection of the environment is an elusive goal. Individuals possess the right to exploit their property as they wish with limited accountability for their actions. The freedom given to the pursuit of individual gain repeatedly produces outcomes that are individually rational, in that they may maximize the return to one's property, but socially irrational. This is the rationale for a system that allows coal-fired power plants in Ohio to continue to operate even though their emissions are polluting the air in the backwoods of Maine.

The existence of individual private property is responsible for the immense inequalities of wealth and income in the world today. Small numbers of individuals hold large amounts of property, whether it takes the form of land, factories, mines, or intellectual ideas. Billions, on the other hand, possess no property of any sort and are forced to subsist on wages earned in the market for labor, which in the last century has increasingly turned against them. Population growth in the twentieth century created a vast army of unemployed, and more recently, globalization of manufacturing has pitted the laborers of the world in competition against each other, with the "winner" being the country that can force its citizens to accept the lowest payment for their work.

Given the ills that can be traced to individual private property, it is surprising that in the last few decades there has been virtually no questioning of the institution, but rather a strengthening of it. In part, this is due to the ascendance of neoliberal economic thought, for which the protection of individual property rights is central. The other part of the story is the failure and collapse of the Soviet Union, which for several decades had represented an alternative form of ownership. Since the disappearance of this "competitor," defenders of individual private property have proclaimed that no other system is possible. But this conclusion is premature. A wealth of historical evidence reveals that human beings have constructed alternative systems of property that are far more egalitarian, superior in their treatment of the environment, and formed the basis of successful and resilient communities.

NEOLIBERALISM AND PRIVATE PROPERTY

Neoliberalism—the dominant economic policy of our times—presumes that individuals will maximize their well-being and the well-being of society if they are allowed to freely buy and sell their goods in markets. The neoliberal economic policy package consists of clearly defined individual property rights and the elimination of any restrictions on the operation of markets both within and between countries, i.e., the so-called system of free trade (so-called because it is in fact highly managed by governments, international institutions, and corporations). Trade, in particular, has received much attention in the last decade as the global trading system has been remade by the North American Free Trade Agreement (NAFTA) and the World Trade Organization (WTO).

The push for free trade has not gone unchallenged, however. Neoliberalism's opponents argue that global free trade gives free rein to corporations whose power far outweighs that of laborers, consumers, and other actors in the international marketplace. Indeed, the economic power of corporations is greater than that of many governments. In a revealing statistic, the total revenues of ExxonMobil are greater than the GDPs of many countries.

Critics of neoliberalism also observe that restrictions on the market may be indispensable for attaining goals such as environmental protection and economic equality. And since at least the 1960s, critics of the World Bank and International Monetary Fund have argued that free trade will not lead to economic development in the world's poorest nations. Historically, industrialization has not emerged from policies of free trade, but from heavy state intervention in the market, as the examples of Great Britain in the eighteenth century, Germany and the United States in the nineteenth, and Japan and South Korea in the twentieth amply testify.

As should be apparent, opponents of neoliberalism have tended to focus on the failings of the market. These correctives to the prevailing market enthusiasm are indispensable, but to focus solely on the market in an evaluation of neoliberalism is insufficient. In my opinion, the market is for the most part a reasonable method for setting prices and distributing goods among people. Far more dangerous than neoliberalism's embrace of the market is its commitment to individual property rights, which it sees as an inextricable part of a market economy. A market economy, however, can function with a variety of ownership forms. Issues of property rights and ownership have received far less attention from environmentalists, representatives of labor, and others who have been active in the anti-globalization movement.

Neoliberalism sees individual property rights as absolutely essential for economic efficiency. Only with private property will economic actors be fully rewarded for their efforts, as a return on their investment, and thereby willing to put forward maximal effort. And only with private property does economic progress occur, as the prospect of future profits leads to innovation and technological change. Not surprisingly, the neoliberal calculation of efficiency does not take into account the social damage and environmental destruction that private property leaves in its wake. Nor does neoliberalism seriously inquire into the links between other systems of ownership and technological innovation. Despite these gaps in the neoliberal analysis, however, it has concluded that the attempt to explore alternatives to private property are unrealistic and utopian. Therefore, neoliberalism is not simply an ideology of the market, but also an ideology of individual private property.

Neoliberal views on property have received little critical attention in part because property itself is not an issue that is up for discussion these days. In the middle decades of the twentieth century there was at least some debate on competing models of ownership, in part inspired by the example of the Soviet Union, which represented an alternative, that of state property. But even decades before the collapse of communism, it was obvious that state owner-ship of property held few attractions. It was problematic on political grounds as it gave overwhelming power to the state and its officials. It was also undesirable on environmental grounds, and the ecological disasters of the Soviet Union, China, and other commu-nist societies have made clear that state property was no better than individual property when it came to preserving the natural world. In fact, it may have been even worse, given the long roster of ecological catastrophes under Soviet rule, including the near nuclear meltdown at Chernobyl, the pollution of the Black Sea, and the desiccation of the Aral Sea and subsequent desertification in the surrounding area. The Soviet state was far more concerned with promoting rapid economic growth and building up its military capabilities than promoting ecologically sustainable forms of pro-duction. And state dominance of economic decisions gave insuffi-cient power to local communities, which would have made different choices because they were the ones that bore the consequences of living with environmental degradation.

In this post-Soviet political context, in which there is little con-sideration of ownership of economic assets, neoliberalism has de-clared that there is no alternative to individual private property. We must reject this claim. Private property certainly has a role to play in any economy, but it must be combined with communal forms of ownership for the greater benefit of individuals, communities, and the natural world.

PROPERTY AND COMMUNITY

Property rights are not natural, but are agreed upon. They are social conventions. If we peer into the past, we learn that in many places, a variety of things that we would not consider property to-day were bartered, sold, and passed on to one's heirs. We would also discover that a variety of ownership forms coexisted comfortably. For example, in India, the Netherlands, France, England, China, and many other places, different conceptions of property, including in-dividual, common, and other communal forms, were applied to dif-ferent types of assets. By contrast, today there is a tendency to believe that one form of ownership, whether it is individual or public, should be universal.

Studying the past also reveals that property has not always been polarized into something that was either owned by a single individual or by everyone in common. It is this polarized view that sees common ownership as the only alternative to individual ownership and produced recent calls for protecting much of the natural world as a global commons. This is a worthy goal as vast parts of the natural world are part of the common inheritance of humankind. But we must go beyond the dichotomy of individual versus common if we are to invent forms of property that meet the needs of humans in their interactions with nature today.

The polarization of ownership into individual versus common has its origins in political and economic developments in the European past. Before the nineteenth century, peasants in Europe possessed a variety of communal rights in property. One type of right was the common field system under which the village community as a whole took decisions about how lands were to be cultivated. A second was common rights to meadows, ponds, and woods, which ensured that all community members had access to land that provided pasturage for animals, timber, fish, and game animals, and other valuable resources. Beginning in the fifteenth century, these communal rights came under attack as wealthy and powerful landlords sought to assert their private ownership and establish their individual say over the use of their property. Over time, many peasants became propertyless as they lost control of their fields and their access to the commons. Drawing lessons from this economic transformation, European observers came to see property as something that was owned either individually or communally, with these forms of ownership incompatible and in conflict. And much thinking on property has followed this framework.

If we widen our lens, however, and consider conceptions of property that existed outside Europe, we will discover that individual or common property are not our only options. The Indian subcontinent is a rich source for alternative forms of property holding, particularly in relationship to community. Indeed, before the establishment of British rule in the eighteenth century, Indians had developed a highly diversified and sophisticated set of property rights.

The subcontinent was certainly home to common property, which as in Europe consisted of woods and uncultivated areas around villages. These commons were essential for the survival of vast numbers of people as they provided timber, land for the pasture of animals, and essential products such as wax and medicinal plants. From the nineteenth century, these common areas began to shrink in size. A hunger for cultivable land, which was driven by a growing population, led to a felling of forests and the reclaiming of land that had formerly lain uncultivated. And the British colonial state began to restrict access to forests in order to regulate the large-scale commercial exploitation of timber and other forest resources. Today in India, for all intents and purposes, common land exists largely in the realm of historical memory.

The Indian subcontinent also possessed property rights that were held individually. This form was typically given to an individual by a superior, either a king or even a god, and usually took the form of a grant with a title to support the right of ownership. A variety of things were considered individual property, including political offices and revenue collection rights. We would think of neither of these as property today, which shows that even the definition of what constitutes property varies from place and time.

There was also a third category of property, which had no counterpart in Europe. This form of property combined both individual and communal features, and I will call it property as share. Land, the most valuable economic resource in precolonial India, came under this designation. Before the coming of British rule, land in India was not something that was owned, transferred, or sold. In other words, there was no property in land. What was owned, however, was a right to a share of the harvest from a particular piece of land. This way of thinking about property is reflected in a Tamil (a South Indian language) word for property, *pangu,* which may be translated as "share." Individuals owned these rights to a share of the output, and *pangus* had been bought and sold since at least the twelfth century. Therefore, at their core, this form of property was individually held.

Yet, as is implied by the term share, these individual property rights existed in the context of a community of individuals, among whom the output was divided. Some information on how this worked has been preserved for villages near Chennai (Madras) in southern India. At harvest time, rice, the major crop, was cut and stored in a central place within each village. Once the size of the harvest had been estimated, a division of it commenced. Members of the community were given a fraction of the output, which was in proportion to the size of each individual's property holding. Portions of the harvest were also kept aside for the financing of routine maintenance on water tanks and channels, which supplied the water that was indispensable for growing rice, and for *pitchay,* or poor relief.

Therefore, in contrast to the European tradition of property in which individual and community rights were seen as opposed to each other, property as share in South India possessed both individual and communal elements. For an individual to exercise a claim to property, he or she had to be a member of a community and work cooperatively within it. And working together, the agricultural community took responsibility for organizing cultivation, maintaining the production system, and investing in its improvement.

In this framework of property rights, the interests of individuals were consistent with those of the community because the only way to increase the size of an individual's share was by increasing the output of the community as a whole. Similarly, the value and viability of an individual's property was dependent upon the value and viability of the community property. Therefore, there were enormous communal pressures to organize agricultural production in ways that were consistent with the long-run environmental sustainability of the land. Property rights as share had existed in parts of South India since at least the tenth century, with some fields cultivated continuously for nearly a thousand years. Individual private property in land was introduced under British rule in the early nineteenth century and in less than a century rice-growing lands were exhausted and ecologically played out. This decline may be traced in part to the removal of the communal element in property, which had acted as a check on the overexploitation of the soil.

When individual private property was introduced into South India, a small group of Indians who the British identified as the landowners quickly monopolized the ownership of land. As a consequence, large numbers of South Indians were stripped of property, which they had possessed in the former share system, and were now forced to earn their keep as wage laborers. For these individuals, the loss of their property meant a loss of economic security as they were no longer guaranteed a share of the harvest, but rather were forced to receive wages, which were set by the market. And over the course of the nineteenth and twentieth centuries, the market for labor increasingly turned in favor of employers. The outcome was growing poverty in South India and a worsening in the conditions of life for ordinary people. Far from being a land of "eternal poverty," as Western conceptions often have it, poverty arose in South India in the aftermath of the shift to a regime of private property, which was further reinforced by population growth in the twentieth century. According to one calculation, the daily income of a man working in rice cultivation was *three times higher in 1795 than in 1976*. The ownership of property, therefore, had been essential to well-being in South India.

There are examples of property as share from Europe as well. Between the thirteenth and nineteenth centuries, access to common lands in the eastern Netherlands was allocated upon the basis of shares, which were considered forms of property. In one village, for example, a holder of a share was entitled to graze ten beasts and four horses in the common land as well as fell one beech tree and one birch tree every year. These shares were bought and sold, with the restriction that outsiders to the village could not purchase them, and they were also passed on to children along with holdings of land. At an annual meeting, which all owners of shares attended, the extent of exploitation of the common for the ensuing year was determined and fines imposed upon villagers who had violated the laws of the commons. This is a system of property in share that closely resembles that of South India.

The South Indian and Dutch property-as-share system centered upon land, but there are parallels to its mixture of individual and communal elements in other places and in other things. One of the most striking was found in eighteenth-century England, where skilled craftsmen viewed the knowledge that they possessed as a form of property, or a "property in skill." Although this was a property that was possessed by individuals, these craftsmen were very careful to argue that it was situated in the context of a community. As one historian has put it: "Even though [skill] was in part an inheritable property which fathers could pass on to sons, what was inherited was rather the use right to be exercised within the regulations and constraints imposed by the trade."

RECONCEPTUALIZING COMMUNITY: FROM AN IDEA TO MATERIAL REALITY

For several decades, a growing chorus of voices has been calling for the rebuilding of our communities. However, reviving the social connections that animated humans in the past has not proven to be an easy task. This failure stems in part from a serious misreading of where community strength and resilience came from in earlier periods. There is a tendency in current discussion to see community as having its basis in feelings of affection, which are cemented through acts of reciprocity and social conviviality. On this reading, to develop stronger communities, we must simply be less selfish, develop a more communitarian ethos, and stop "bowling alone," in Robert Putnam's evocative phrase. Communities in the past, however, did not rest mainly on a more communal state of mind or on emotional bonds that were formed by community events. Historically, the resilience of communities came from the fact that they were rooted in material interests and economic institutions. In the case of India, it was communal and localized control over property that gave structure, strength, and meaning to communities. Therefore, to revitalize our communities and conserve natural resources, we must reassert these principles of ownership.

A PROPOSAL FOR THE PRESENT

The property as share system was applied to land in India. Nevertheless, the principles that guided the system may be applied more generally, and are particularly appropriate for industrial property and natural resources. As I see it, the central principles of this form of ownership are the following: First, property rights in a share of the output rather than in the factory or mine; second, to maintain a balance between individual and communal components in the property right, sale of the property is permitted but with the approval of other owners of shares; finally, to preserve a link between production decisions and the environment, all owners must live in the locality where the economic operation is located.

To take the example of a manufacturing enterprise, all workers, from cleaning staff to managers, would be vested with property that would consist of a share in the output of the operation. All owners would be required to contribute to the productive activity of the enterprise with provisions for retirement. Property could be transferred, either passed on to one's heirs or sold, of course with communal approval and with the proviso that the new owner of the property would participate in the manufacturing enterprise, i.e., there would be no absentee ownership of shares. Similarly, in the case of natural resources such as minerals, forests, and farmland, there would not be individual ownership of the resources themselves, but individuals who participated in their exploitation would be given a property right in a share of the product.

The requirement that all owners have to reside in the same locality as the factory or mine helps to ensure that ecologically sustainable practices will be adopted. If those who create pollution have to live with its consequences not only at work but also at home, in streets, and at schools, there are powerful incentives to choose clean production techniques. Indeed, it is the ubiquity of chemical toxins from decades of herbicide and pesticide use that is leading many small U.S. farmers to shift to organic farming.

The right to pass on property to heirs creates a link between present and future. With a system of property as share, this link would exist not only for one or two individuals, which is typically the case where property is concentrated in a few hands, but for everyone involved in the economic operation. Therefore, there would be enormous pressures to preserve the value and viability of the operation for future generations. Environmental sustainability of the production process would become crucial for the property right to retain its value.

If property rights were reconceived as a share of output, we would also have a more egalitarian society. The extremes of wealth and poverty that exist in the United States today would be narrowed as property would be diffused widely among the population.

A system of property as share would also create a more meaningful democracy. Democratic principles would no longer be restricted to the political sphere and would enter the economy, daily life, and work. In a world of property as share, all producers would possess the right to participate in the decisions that shape their fate at work, which after all will also shape the fate of their property. A more democratic workplace would be one result.

To reshape property in these ways would bring manifold social benefits. However, many things are not amenable to being divided into shares. In small family-operated economic enterprises, such as family farms or shops, private property would continue to prevail. So there would certainly be room for individual private property, but it would be one among other forms of ownership. At the same time, it is essential to reinvigorate the idea of common property, especially in resources that may not fall under the ownership of individual communities, including air, water (both surface and underground), and fisheries. These must be seen as common resources of humanity—as a new global commons—with new institutions to regulate their exploitation in order to preserve them. Despite common misconceptions about the infeasibility of restricting use of common resources—the famous tragedy of the commons—there are plenty of examples of successful institutional setups that have preserved common resources. Filipino irrigation systems, Turkish fisheries, and Alpine forests have all been sustainably managed for centuries.

I recognize that it will require sustained political struggle to transform our system of ownership in the ways I have described. Nevertheless, to begin the battle we must have a goal and a vision. The purpose of this paper has been to commence that battle by putting property rights back on the political agenda. We must reject the neoliberal claim that individual private property is the only viable form of property. There are alternatives—far superior ones. We must reclaim the property rights that only a few centuries ago were distributed so widely. Only then can the welfare of individuals be reconciled with that of communities and the planet.

ANOTHER WORLD IS POSSIBLE:
NEW RULES FOR THE GLOBAL ECONOMY

Sarah Anderson and John Cavanagh

Over the past half decade, public debate about the global economy has broken wide open. Exciting new discussions concerning trade, financial institutions, and development policies are taking place around the world. How can international institutions be revamped to serve the needs of people? Can we restructure our rules of trade, finance, and North/South interactions to close the dramatic gap between rich and poor? How can we ensure an ecologically sustainable future? The debate is that basic.

This chapter briefly recaps the problems with the status quo and then summarizes some of the best new thinking and action for creating a new global trade and finance system. This new vision, fueled by expanding citizen movements for global justice, is still developing and is not ours alone. We belong to a number of networks of academics, citizen activists, and enlightened business and government leaders who are working across borders and across sectors to build new models for the global economy.

FIGHTING FOR THE STATUS QUO

The governing formula of the past half century has been that unfettered trade and investment will bring prosperity, which will bring democracy. This view has guided the declarations of U.S. presidents from Truman to George W. Bush as well as the policy pronouncements of political leaders the world over, particularly since the early 1980s. It has also been the mantra of the three main public international economic bodies—the World Bank, the International Monetary Fund, and the World Trade Organization. The argument is that nations must be competitive in a global economy and that competitiveness requires governments to reduce barriers to trade and provide the most favorable possible climate for foreign investment. Governments should enhance the ability of corporations to move their factories, money, and products around the globe more quickly and with fewer regulatory impediments. By increasing corporate mobility, such policies will enable international firms to become more efficient and profitable. The profits will eventually trickle down to the rest of us. Or so the argument goes.

As public criticism of this free-market approach to the global economy has grown in recent years, defenders have fought hard to preserve the status quo. In a July 2001 speech at the World Bank, President George W. Bush angrily lashed back at the critics, charging that "those who protest free trade are no friends of the poor. Those who protest free trade seek to deny them their best hope for escaping poverty." In the wartime climate of the fall of 2001, the Bush Administration added a new argument: that free trade is necessary to fight terrorism. Top officials, including Secretary of State Colin Powell, based this claim on the assumption that globalization contributes to economic prosperity, which in turn promotes global security.

Meanwhile, a growing body of evidence counters these arguments. And the criticism is coming not just from the protestors in the streets, but from a number of economists at institutions that are far from hotbeds of radicalism. Harvard University economist Dani Rodrik claims there is no relationship between an absence of trade barriers and economic prosperity. He notes that many of the more successful economies, including the so-called East Asian Tigers, undertook trade liberalization only very gradually and only after building up domestic markets. Rodrik is particularly critical of rapid liberalizing of capital markets that, in his opinion, contributed to recent financial crises in Mexico, Thailand, and Turkey.

Joseph Stiglitz, a former World Bank chief economist who recently received the Nobel Prize in Economics, has also condemned the IMF for pushing countries throughout the developing world to liberalize their capital and financial markets prematurely. Likewise, the Inter-American Development Bank has argued that "liberalization of financial markets has tended to increase poverty and inequality in Latin America," pointing out that gains in economic growth during the past decade were swept away by rising inequality in every country in the region for which data are available. At the global level, the United Nations studied seventy-seven countries that are home to 82 percent of the world's population and found only sixteen that had experienced a narrowing of the gap between rich and poor.

Some promoters of corporate globalization maintain that rising inequality is not grounds for concern so long as the overall economic pie is expanding. This idea is hotly contested by others who argue that growth based on exploitation can hardly be considered a sign of success. However, even if one were to accept this narrow standard, the record is mixed, at best. A study by the Center for Economic Policy and Research, for example, reveals that most developing countries had faster growth rates during the 1960s and 1970s than they did during the two most recent decades. This means that during the period when the World Bank and IMF were most vociferously pressing globalization policies, the economic pies of most countries in the developing world actually grew more slowly than before. Among the poorest countries, per capita GDP growth dropped from 1.9 percent annually from 1960 to 1980, to 0.5 percent per year from 1980 to 2000. For the middle group, there was a sharp decline from an annual per capita growth rate of 3.6 percent to just less than 1 percent.

Meanwhile, globalization policies have also failed to help the developing world reduce its debt burden. Between 1980 and 2000, the foreign debt of all developing countries increased more than fourfold, from $586.7 billion to more than $2.5 trillion. As a result, governments must divert money away from feeding the hungry and treating the sick in order to meet debt payments. The debt burden also places strong pressure on governments to try to attract foreign investment and promote exports. These efforts frequently threaten both the environment and workers. In Mexico, Indonesia, and elsewhere, this has meant busting unions. In Chile, the Philippines, and other natural resource–rich nations, debt repayment has involved cutting down forest for timber exports or plantation expansion, depleting fishing stocks, and expanding open-pit mines.

The Case of NAFTA

The North American Free Trade Agreement, initiated in January 1994, goes farther than any agreement in the world to promote understrained free trade and investment between countries with extreme gaps in living standards. During the intense fight over NAFTA in the U.S. Congress in 1993, promoters argued that the trade pact would lift living standards and strengthen protections for workers' rights and the environment in all three countries (Canada, the United States, and Mexico).

Eight years later, the reality is quite different. Mexico has experienced a rapid increase in net foreign investment, from $4.4 billion in 1993 to $11.8 billion in 1999. Not surprisingly, the country's exports to the United States have also shot up, from about $50 billion in 1994 to $136 billion in 2000. However, average Mexicans have seen few of the benefits of this globalization. According to the World Bank, the percentage of Mexicans living in poverty has grown since NAFTA began, from about 51 percent in 1994 to more than 58 percent in 1998.

Although NAFTA's labor-side agreement was promoted as a means of preventing egregious violations of worker rights, none of the more than twenty complaints filed under the treaty's dispute resolution mechanism have resulted in anything more than consultations between the NAFTA governments. Meanwhile, Mexican workers who struggle to fight for better wages typically face severe repression. In one particularly brutal example, U.S.-based Duro Manufacturing colluded with Mexican officials in 2001 to squash an attempt by workers in a Rio Bravo, Mexico, factory to form an independent union. As the workers demonstrated peacefully in front of the plant, local police swept in and beat them. Later, arsonists torched the house of the workers' leader. When the time came for the union election, the Mexican government reneged on promises to allow a secret ballot, forcing terrified workers to vote in front of management, with armed thugs hovering nearby.

The impacts of NAFTA on Mexico's environment have also been devastating. The explosion of industrial development in the

U.S.-Mexico border region has not coincided with sufficient investment in environmental infrastracture to handle the increase in pollution from both industry and the growing population. A recent study by Tufts University indicated that pollution from Mexican manufacturers nearly doubled between 1994 and 1999, a reversal of trends during the previous five-year period.

In the United States, NAFTA promoters argued that the trade pact would create more and better jobs, based on their prediction that the U.S. trade surplus with Mexico would increase. Just the opposite has occurred. Whereas the United States had a small trade surplus with Mexico before NAFTA, it now has a huge trade deficit. The exact number of U.S. workers affected by NAFTA is difficult to calculate. However, as of November 2001, about 367,000 U.S. workers had qualified for a special NAFTA retraining program for people who lose their jobs because their employer moved production to Mexico or Canada or was hurt by import competition from those countries. U.S. communities on the border with Mexico have been especially hard hit, as thousands of manufacturing jobs have been shifted to the Mexican side. Moreover, the industries with the most NAFTA-related job losses, apparel and electronics, are major employers of women and people of color.

The impact on U.S. wages is perhaps even more striking. After several years of stagnation, U.S. real wages increased at the end of the 1990s, but only slightly, despite high corporate profits and record low unemployment. The value of the median wage in 2000 was still below 1973 levels. Meanwhile, executive pay jumped 571 percent between 1990 and 2000. In a study of hundreds of union organizing campaigns in "mobile industries" (manufacturing, communications, and wholesale distribution), Cornell professor Kate Bronfenbrenner found that in 68 percent of the cases, management fought the union by threatening to close the plant and move to Mexico or other low-wage countries. Even some free trade proponents admit that globalization pressures are a major part of the explanation for the negligible wage gains in a time of prosperity.

All three NAFTA countries have been affected by the agreement's investor protections, which make public interest regulations vulnerable to expensive corporate lawsuits. For example, the Mexican government was forced to pay nearly $17 million in damages to a California hazardous waste treatment firm that sued after a Mexican municipality denied it a permit to operate a facility on public health and environmental grounds. The U.S. government has ignored demands that these rules not be used as a model for future agreements. In the recently released draft of the Free Trade Area of the Americas, the NAFTA investment language is repeated virtually verbatim.

THE GROWING BACKLASH

In recent years there has been an explosion of demonstrations against the WTO, the World Bank, the IMF, and the corporate globalization agenda more generally. This growing movement is unlike any other in the breadth of its composition and in its cross-border linkages. Labor, environmental, human rights, consumer, family farm, student, and faith-based groups have joined forces with counterparts in many areas of the world. Although the mainstream media has often given the impression that these diverse organizations are hopelessly fragmented there is actually a great deal of unity. The citizen backlash everywhere decries the excessive political and economic power of large global corporations. It believes that the main institutions and policies that govern the global economy favor narrow private interests over the common public good and that these institutions should be drastically reformed or eliminated. And it holds that the current system of corporate globalization, rooted in what George Soros has called "market fundamentalism" is not inevitable.

There is also considerable consensus around what members of this movement are proposing. Generally, these groups advocate replacing market fundamentalism with the principles of living democracy, universal rights, ecological sustainability, and fairness. They favor stronger checks on corporate power at the local, national, and global levels. They believe that certain goods and services, such as water and other parts of the global commons, should not be subject to global market forces or agreements. They feel that other aspects of the global economy, such as speculative capital flows, should be slowed down. Finally, they advocate changing rules and institutions so that economic exchanges support healthy communities, dignified work, and a clean environment.

One showcase for the wealth of ideas around a positive alternative agenda is the World Social Forum, where thousands of international activists converge to create a counter to the World Economic Forum, the annual elite gathering of corporate and government supporters of globalization. The first annual World Social Forum in Porto Alegre, Brazil, in January 2001 carried the slogan "Another World Is Possible" to encourage activists to "think big" about their ideal vision for the global economy. The following sections of this chapter focus in detail on two citizen efforts that the authors have been involved in—one in the Western Hemisphere and one on a global plane—that offer constructive paths to another world.

HEMISPHERIC SOCIAL ALLIANCE

The Hemispheric Social Alliance, a network of labor organizations and citizens coalitions representing close to 50-million people in the Western Hemisphere, has sponsored a multiyear collaborative process among scholars and activists to articulate a positive alternative to the free-trade model. These views are summed up in a document entitled "Alternatives for the Americas," which presents a detailed alternative to the Free Trade Area of the Americas, the hemispheric trade pact scheduled for completion in 2005. After first releasing it in 1998, the Alliance continues to gather input on the document and periodically releases new versions.

Our involvement in this process is through the Alliance for Responsible Trade (ART), a U.S.-based coalition that is a member of the Hemispheric Social Alliance. We have drawn heavily on our experiences working with ART in the early 1990s to forge alternatives to NAFTA along with counterparts in Mexico and Canada. Like these early initiatives, the basic guiding principle of the "Alternatives for the Americas" document is that trade and investment should not be ends in themselves, but rather the instruments for achieving just and sustainable development. It argues that a common human rights agenda should form the overall framework for all hemispheric policies and include mechanisms and institutions to ensure full implementation and enforcement. This agenda should promote the broadest definition of human rights, covering civil, political, economic, social, cultural, and environmental rights, gender equity, and rights relating to indigenous peoples and communities.

A key component of the agenda set forth in "Alternatives for the Americas" is a strong call to ease the debt that has a stranglehold on most countries in the region. In 2000, countries in Latin America and the Caribbean had a total debt burden of $809 billion, more than three times as high as in 1980, despite having made more than a trillion dollars in debt payments during that period. In 2000 alone, the region paid $167 billion in principal and interest on the debt—money that otherwise could have been spent on basic services. Recently, the debt toll has been most evident in Argentina. On the brink of default throughout 2000, the government sought emergency lending from the IMF that was conditioned on severe austerity measures. Squeezed by deep cuts in public salaries and services and an unemployment rate of more than 20 percent, the populace revolted and more than twenty-five people were killed during riots. By early 2002, the country descended into total political and economic collapse.

Although critics argue that debt relief is unrelated to trade policy, measures to address this burden would actually build on the precedent set by the European Union. There, richer countries funneled development aid into Spain, Portugal, Greece, and Ireland to lift up their living standards during early stages of integration. In order to level the playing field in the Western Hemisphere, the "Alternatives for the Americas" calls for the 100 percent cancellation of debts of low-income countries. In Latin America and the Caribbean, this would involve Bolivia, Guyana, Honduras, Nicaragua, Haiti, Jamaica, and Peru. With regard to middle-income countries (e.g., Brazil, Argentina, Bolivia, Mexico, Paraguay), it calls for the creation of an arbitration panel to assess and work out the debts, along the lines of the arbitration mechanism established through U.S. bankruptcy law. (In late 2001, the IMF dropped its longstanding opposition to such a process, but proposed that it play the role of arbitrator, which, since the IMF is itself a creditor, would be unacceptable.)

The document also offers creative proposals for preventing a "race to the bottom" in wages (i.e., a process in which countries compete for foreign investment by offering lower and lower wages). It recommends that countries be required to comply with core worker rights, as defined by the International Labor Organization (ILO). These include the right to freedom of association and collective bargaining and bans on slave labor, child labor, and discrimination. The document proposes that unions and nongovernmental organizations be allowed to file complaints with the ILO regarding violations of ILO core standards. The ILO would then investigate to verify whether or not the conventions had in fact been violated. At a second stage, the ILO would formulate recommendations to the country to assist it in complying with the conventions that have not been respected. If this second stage were unsuccessful, then the perpetrator would be sanctioned.

We feel that one of the most innovative aspects of the proposal is that it would allow sanctions to be applied directly against a corporate violator, instead of just against a country. (Currently, neither the NAFTA labor-side agreement nor any other international agreement allows for punishments against corporations for violating international labor rights). The proposed punishment would be the loss of the benefits of the trade agreement. Thus, if a particular company was found to have violated the right of freedom of association, that particular company's exports would not receive the same preferential treatment granted to all other countries that are members of the agreement. Sanctions against an entire country would only be applied if the country's government were shown to be an active and repeated accomplice in the violation of fundamental workers' rights. This approach is designed to help hold corporations and governments accountable for the most extreme forms of rights violations and protect workers who are struggling to gain their fair share of the benefits of globalization.

Similarly, the document proposes that a race to the bottom in environmental standards should be discouraged by subordinating trade and investment to policies that prioritize sustainable development and environmental protection. Drafted by a Mexican environmental group (with input from around the hemisphere), the goal of this chapter would not be to try to impose rich-country environmental standards on poor nations, but rather to promote adherence to agreements negotiated at the international level. For example, it states that trade rules should not undermine the Basel Convention (which prohibits the export of dangerous waste products from OECD countries to non-OECD countries), the Montreal Protocol on Substances that Deplete the Ozone Layer, and the Kyoto agreements on greenhouse gas emissions.

The environmental chapter of "Alternatives for the Americas" also argues that governments should have the power to channel investment toward environmentally sustainable activities, reject privatization of natural resources, and eliminate policies that subsidize or encourage fossil fuel energy. It criticizes the trend toward establishing environmental regulations on the basis of "risk assessment" (which applies economic cost-benefit analysis to environmental resources). Instead, the document advocates for the use of the "precautionary principle" that, when in doubt, we should take the most environmentally cautious course of action.

On the issue of investment, we have worked closely with groups in Mexico and Canada to draw lessons from the negative experiences with NAFTA's investment rules. Thus, we call for a ban on mechanisms such as that in NAFTA that allow private investors (but not citizens' groups) to sue governments directly over public-interest regulations. Although we make it clear that we are not opposed to all foreign investment, the document would give governments the right to screen out investments that make no net contribution to development, especially speculative capital flows.

On some issues, there remains a lack of consensus. On immigration, for example, some argue that if capital and goods are free to move across borders, then people should be, too. If the EU can have open borders, why can't we? Many Latin American organizations agree with this analysis. Within the United States, on the

other hand, there has been resistance to an open-border policy ✓
from those who argue that an immigrant surge could be detrimen-
tal to people on the bottom end of the U.S. wage scale. However,
there is currently a reinvigorated dialogue on the immigration
question that has created a space for new thinking. One major contribution was an AFL-CIO resolution in 2000 that called for am-
nesty for undocumented workers in the United States and an end to
sanctions against employers who hire them. A reversal of the federation's previous position, the policy has helped bolster the efforts of Latin Americans, including Mexican president Vicente Fox,
to push the U.S. government for a more flexible immigration stance.
While this debate continues, the "Alternatives for the Americas"
lays out consensus points around the need to strengthen enforcement of immigrant rights and to address the root cause of illegal
immigration—the policies that result in economic desperation.

INTERNATIONAL FORUM ON GLOBALIZATION

During the same years that citizen groups were creating opposition and proposals in the Western Hemisphere, the International
Forum on Globalization (IFG) began bringing together leading activists and thinkers from North and South to create large educational events on corporate globalization. In 1999, the IFG formed
a task force on alternatives, and produced an innovative menu of
alternatives that are being tested in regional forums around the
world. Compared to the Hemispheric Social Alliance process
around the FTAA, the IFG project has been far broader, taking on
the whole range of international institutions and regions.

The IFG discovered that in many communities, regions, and
countries, citizen organizations have built creative and workable
social and economic alternatives. In a number of countries, civil society organizations have crafted national processes to build blueprints for more sustainable societies.

- In India in June 1999, thousands of people came together in the state
  of Uttar Pradesh to launch the "Living Democracy Movement" (Jaiv
  Panchayat) in order to, in their own words, "assert legitimate people's

control over ownership of all the biological resources." This movement is an experiment in creating new ways to make the community
the decision-maker on all matters pertaining to biodiversity and its
conservation.

• In Canada over the past half decade the Council of Canadians has led
a process of building a "citizens' agenda" to assert basic rights and
build pressure to reverse the trend of government abandoning its
role as protector of basic rights.

• In Chile in 1997, citizen movements created a process of elaborating,
region by region, a plan to build a "Sustainable Chile." The next year,
similar processes were initiated in Brazil and Uruguay. This momentum to create bottom-up citizens' agendas is now spreading into a
regional exercise for the Southern Cone of Latin America.

ALTERNATIVE PRINCIPLES

As IFG Alternatives task force members gathered citizen
statements and manifestos from around the world, they noticed an
encouraging convergence of groups around an alternative set of
principles for guiding the construction of rules and institutions to
govern economic activity. They also quickly concluded that the current organizing principles of the governing institutions of the
global economy are narrow and serve the few at the expense of the
many and the environment. Economic growth has been the central
goal of the International Monetary Fund (IMF), the World Bank, and
the General Agreement on Tariffs and Trade (GATT), as well as its
successor, the WTO. The expansion of international trade and investment flows has been viewed as an end in itself. Instead, the IFG
asserts that healthy societies are rooted in an alternative set of
core principles, including the following:

New Democracy. The rallying cry of the amazing diversity of
organizations that converged in Seattle in late 1999 was the simple
word "democracy." Democracy flourishes when people organize
to protect their communities and rights and hold their elected officials accountable. "Democracy" is equated in the minds of many

with elections alone, but this movement encompasses new forms of more active democracy. In some countries, primarily in the South, like Indonesia and the Philippines, these movements focus on winning community control over natural resources. In other countries, mainly in the North, movements are striving to refocus government agendas on a citizens' agenda of rights.

Subsidiarity. Economic globalization entails first and foremost the loss of local control over economic activity. Yet, well over half of the people on earth still survive through local, community-based activities: small scale farming, local markets, local production for local consumption. This has enabled them to remain largely in control of their economic and food security, while also maintaining the viability of local communities and cultures. Economic globalization is rapidly dismantling these local economies, strongly favoring economies based on export, with global corporations in control. Right now, in no country is this happening more rapidly than in China, as it has rushed to adopt the investment and trade-opening rules of the WTO. This brings destruction of local livelihoods, local jobs, and community self-reliance. It is therefore necessary to reverse direction and create new rules and structures that consciously favor the local and follow the principle of subsidiarity, i.e., whatever decisions and activities can be undertaken locally should be. Whatever power can reside at the local level should reside there. Only when additional activity is required that cannot be satisfied locally should power and activity move to the next higher level: region, nation, and finally the world.

Ecological Sustainability. Economic globalization is intrinsically harmful to the environment, as it is based on ever-increasing consumption, exploitation of resources, long-distance transport, and waste-disposal problems. Fossil fuel-based trade and energy systems pollute the air and water and have engendered planet-threatening climate change. Export-oriented production systems are especially damaging as they directly increase global transport activity, fossil fuel use, refrigeration, and packaging, while requiring very costly and ecologically damaging new infrastructures: ports, airports, dams, canals, etc. In short, the global economy is substituting unecological activities (industrial farming, autos, chemicals, etc.) for more sustainable ones (self-reliance, low-tech, etc.). Viable alternatives must be rooted in the principle of ecological sustainability.

Human Rights. In 1948, governments of the world came together to adopt the United Nations Universal Declaration on Human Rights, which established certain core rights, such as "a standard of living adequate for ... health and well-being ... , including food, clothing, housing, and medical care, and necessary social services, and the right to security in the event of unemployment." Building on this declaration, governments negotiated two covenants in subsequent decades, one on political and civil rights and another on economic, social, and cultural rights. Citizen groups are asserting that it is the duty of governments to insure these rights, not only the political and civil rights, but also the economic, social, and cultural rights.

Jobs/Livelihood/Employment. The United Nations Universal Declaration on Human Rights affirms everyone's "right to work, to free choice of employment, to just and favorable conditions of work, and to protection against unemployment." It also affirms that everyone "has the right to form and to join trade unions." Accordingly, sustainable societies should both protect the rights of workers and address the livelihood needs of the 30 percent of humanity that has no work or is seriously underemployed. The reversal of globalization policies that displace farmers from their land and fisherfolk from their coastal ecosystems are central to the goal of a world where all can live and work in dignity.

Food Security and Food Safety. Communities and nations are stable and secure when people have enough food, particularly when nations can provide their own food. People also want safe food, a commodity that is increasingly scarce as global agribusiness firms spread chemical- and biotech-intensive agriculture around the world. Any new rules of trade should recognize that food production for local communities ought to be at the top of a hierarchy of values in agriculture. Local self-reliance in food production and the assurance of healthful, safe foods should be considered basic human rights. Shorter distances and reduced reliance on expensive inputs that which must be shipped over long distances are key objectives of a new food system paradigm.

Equity. Greater equity reinforces both democracy and healthy communities. Economic globalization, under the current rules, has widened the gap between rich and poor countries and between rich and poor within most countries. The social dislocation and tension that result have become the greatest threats to peace and security the world over. Reducing the growing gap between rich and poor nations requires first and foremost the cancellation of the illegitimate debts of poor countries. And it requires the replacement of the current institutions of global governance with new ones that include global fairness among their operating principles.

Diversity. A few decades ago, it was still possible to leave home and go somewhere else where the architecture was different, the landscape was different, the language, lifestyle, dress, and values were different. Today, farmers and filmmakers in France and India, and millions of people elsewhere, are protesting to maintain that diversity. Tens of thousands of communities around the world had perfected local resource-management systems that worked. Those systems are now being undermined by corporate-led globalization. Cultural, biological, social, and economic diversity are central to a dignified, interesting, and healthy life.

PROTECTING THE GLOBAL COMMONS

Citizen groups in India, the Philippines, Canada, and elsewhere, as well as many indigenous societies, are asserting that certain goods, services, and activities should be off limits to globalization forces because they involve a global or national commons that needs to be protected. This "commons" include fresh water, air, genes, and seeds, as well as public services that address basic needs, such as public health and education. The IFG Alternatives task force offers four categories for discussion on the policy aspects of the commons and basic rights. The categories are meant to spark a dialogue and are not set out here as final or definitive.

1. Not to be traded because they are pernicious to the environ-

ment or public health, safety, and welfare. This list includes some goods that are currently traded and some that are not. It could include toxic waste, endangered species, nuclear technology and waste, armaments, and genetically modified organisms. Trade institutions like the WTO should not use their power to overturn international agreements that limit or forbid the trade in these areas.

2. *Not to be commodified or traded for commercial profit because they are necessary for human and ecological survival.* This list would encompass the fundamental building blocks of life that are necessary for human and ecological survival. Air, genes, and bulk water belong to the earth and all species; no one has a right to appropriate them or profit from them. All should be declared a public trust to be protected by all levels of government and communities everywhere.

3. *Not to be patented because they are the common inheritance of humanity.* This list would include those areas of life that might be traded under fair-trade rules or are currently traded in traditional communities, but that should not be patented for private corporate use. It would include life-forms that are traded, such as seeds, plants, and animals, and some that should not be traded or patented, such as genes and human genome. It would also include patented drugs, such as generic AIDS drugs, that are sold and traded, but should not be patented for profit.

4. *Not to be subject to current international trade institutions like the WTO because they are the common public rights of all peoples.* This list includes those areas of the commons and basic rights that are fundamental to self-reliance of communities and self-determination of peoples. In some cases, such as agriculture, citizens and their governments would likely encourage trade in these sectors. However, they would always retain the right to set fair-trade conditions in order to maintain domestic control of the sector. In other cases, such as health care, governments might declare these areas public rights and off-limits to privatization altogether. In any case, no international trade agreement based on the current notion of free trade would have the right to overrule the democratically determined definitions in any of these sectors. They would include: food and agriculture; natural resources (except air and

water); culture and heritage; health and education. Also, this category should include goods that are unsafe or unhealthy, such as tobacco; no government should be obligated by a trade agreement to open its market to foreign tobacco products.

NEW INSTITUTIONS

Finally, the IFG makes a strong case that the World Bank, the IMF, and the WTO need to be replaced with new institutions responsible to the United Nations charter. Here are three examples of the kinds of institutions they believe are needed:

An international insolvency court. Debt relief rather than the provision of still more debt is the more appropriate response to the over-indebtedness of low-income countries. A people cannot be both free and in debt. Recommendations to create an international insolvency court (IIC) have come from the United Nations Conference on Trade and Development, the Jubilee 2000 Coalition, and the Canadian government, among others. The IIC would be comprised of a conciliation panel and an arbitration panel. The conciliation panel would facilitate negotiated settlements between debtors and creditors. The IIC can become operational as the World Bank and the regional development banks are decommissioned, with any remaining assets of the decommissioned institutions applied to debt relief.

An international finance organization (IFO) under the mandate and direction of the United Nations. The International Finance Organization would work with UN member countries to achieve and maintain balance and stability in international financial relationships, free national and global finance from the distortions of international debt and debt-based money; promote productive domestic investment and domestic ownership of productive resources; and take such actions as are necessary at the international level to support nations and localities in creating equitable, productive, sustainable livelihoods for all. The IFO would serve as a replacement for the IMF, but with full accountability to the United Nations. Lacking either lending capacity or enforcement powers, its functions would be to maintain a central data base on international accounts, flag problem situations, and facilitate negotiations among countries to correct imbalances. Its charter would mandate that it favor human, community, and environmental interests over the interests of global corporations and financiers in all its activities.

An organization for corporate accountability (OCA) under the mandate and direction of the United Nations. Next to financial speculation, the greatest economic threat to the well-being of people and planet is the growth and abuse of unchecked corporate power. Today, the sales of the world's two hundred largest corporations are greater than a quarter of the world's measured economic activity. There are virtually no mechanisms in place at the international level for dealing with this threat, and the Bretton Woods institutions regularly seek to block corrective actions at national levels. Corporations, many of which are larger than most states, move freely around the world buying off politicians and playing states, communities, and people against one another in a competition for the jobs, finance, and access to technology that corporations control. Eliminating the Bretton Woods institutions, rescinding structural adjustment agreements, and canceling provisions in international trade agreements that place the interests of global corporations and financiers ahead of human interests will be major steps toward restoring the right and responsibility of governments to hold corporations accountable to public interests.

CONCLUSION

For too long, most governments' policies toward the global economy have been held hostage to a rigid free-trade and investment formula. Now a growing global citizens' backlash made up of unions, environmental groups, and other organizations is demanding a place at the negotiating table to craft new rules that steer the benefits of economic activity to the majority, not the minority, and to ensure that our planet will be preserved in the twenty-first century.

While these efforts face tremendous obstacles, much progress that has been made in recent years. In 1998, international activists (particularly in Canada and France) spearheaded the defeat of the

Multilateral Agreement on Investment, which would have expanded NAFTA-style investor protections to many other countries. In the late 1990s, religious and other activists succeeded in pushing several rich country governments to cancel poor country debts and got the World Bank and IMF to acknowledge the need for more debt relief. The late 1990s also saw an explosion of U.S. student activism against sweatshops that resulted in university codes of conduct for purchasing campus gear. In the developing world, workers, peasants, and others have fought successfully in some countries against particularly onerous World Bank and IMF policies. In 2001, international activists allied with some governments to push the WTO to ease its intellectual property rights protections in order to give poor countries better access to discounts on AIDS drugs. That year, activists also moved the governments of Canada and some European countries to promote the idea of a tax to curb speculative financial flows.

In the United States, activists fought a valiant battle in late 2001 in the House of Representatives to beat back efforts to pass "fast-track" trade authority, which allows the executive branch to negotiate new trade agreements that Congress must vote up or down with no changes. In the end, they lost by only one vote and only after the White House cut last-minute deals and turned the issue into a wartime test of patriotism. The divisiveness of the issue means that passage of future trade deals is highly uncertain.

Of course most of these battles were defensive. And unfortunately activists will have to continue to work to put out the fires of an expanded corporate globalization agenda. However, as this vibrant movement continues to expand and mature, it is increasingly important that we move beyond defensive campaigns. Citizen groups are now equipped with agendas to build sustainable societies and the global rules that offer them space to succeed. Another world is possible.

TIMESHIFTING

Stephan Rechtschaffen

Do you have enough time in your life?

When I ask this question in my seminars on wellness and time, only one or two people generally say yes—out of a class of fifty. When I recently asked this question at a Fortune 100 company, not one of the one thousand people present raised a hand to say yes. Even more surprisingly, when I asked that of sixteen prison inmates with minimum sentences of twenty-five years to life, not one of them raised a hand. So heavily scheduled and regimented were the inmates' hours that even "doing time," they had no time.

In a sense, we are all imprisoned—by the perception that time is a scarce and limited resource. Since the Industrial Age, time has become a measure of our productivity. We abhor the idea of "wasting time" and live with the belief that "there's not a minute to spare." Manacled to our watches, we find ourselves in the fast lane all the time, perpetually rushing from one activity to the next. We're surrounded by labor- and time-saving devices undreamed of by past generations, yet few of us have any time to spare.

Amid our success and ability to produce and do more, there is a nagging sense that modern life, with all its material advantages, is more stressful than ever. In our haste to speed up our lives, to get more done, to have more experiences, something has gone drastically wrong. We have more "stuff" than any people who have lived before us—but this hasn't made us satisfied.

Caught by the dysrhythms and dysfunction of our time, we rush through life without experiencing it. As a bumper sticker states: HAVING A GOOD TIME, WISH I WERE HERE.

Thomas Moore starts his book *Care of the Soul* with the statement: "The great malady of the twentieth-century, implicated in all of our troubles and affecting us individually and socially, is—loss of soul." We could equate the words time and soul. Without taking time to really be present in our lives, to find the time to slow down, we find ourselves rushing through everything we do, riding on the surface of things. Thus we lack satisfaction in our lives, we miss a sense of connection with each other, our children, and our community. Yet there is a way to change this. As we create more time in our lives, we have the opportunity to connect more deeply with our purpose and to regain the sense of soulfulness that brings meaning and contentment to our lives. The path to greater balance is both personal and societal. Significant change is possible simply through a transformation of our perception and experience of time. Truly durable change will also be hastened by policies and practices that provide new options for daily work patterns.

BEYOND LINEAR TIME

We can choose to change our relationship with time. Though we feel caught in the whirlwind of life speeding by and at the effect of time pressure, we can learn to live at ease in the flow of time. Even as we are active and rushed we can open moments when it all slows down—we can learn to create more "time freedom" amid the frenzy of modern life.

First, it's important to recognize that we have become habituated to see time as severely limited by definition. To most of us, time means clock time: sixty seconds a minute, sixty minutes an hour, twenty-four hours a day; unchangeable, inexorable clock time.

I believe it is possible to think of time as something else. We can define time as the rhythmic dimension of life. We can become aware that it contains myriad rhythms and that any individual moment can be expanded or contracted under our control. Thus we can make time our friend, not an enemy—and in so doing, fill our lives with happiness and health to a degree most of us cannot even imagine. And most important, as each person does this, it affects the overall health of society—reducing violence and alienation and engendering more caring and a sense of connectedness.

Note that I am not just talking of "slowing down," but of shifting with ease from one rhythmic pattern to another—fully inhabiting each moment, no matter what its speed. We're each aware that different activities have different speeds, learning to adjust our internal rhythm to these speeds enables us to be more easily present in our lives instead of continually resisting what is happening. Our stress results from resisting this moment, our happiness is learning to simply be in each moment just as it is. Real success in modern times is not to emulate the turtle but to move seamlessly and decisively among different rhythms. Timeshifting, as I call it, is like shifting gears on a bicycle: With ten gears, you can easily go up a steep hill, zoom around a curve, cruise down a hill. But many of us only have one gear—the ever-quickening speed of work—and we multitask, eat, talk with our children, even make love in that gear. Instead, one can easily develop the ability to access the moment at a rhythm we consciously choose—and thus produce a rhythmic variability that is associated with greater vitality and health.

We can inhabit the present—just as it is. As the saying goes, the present is the present. Like any gift, it must be opened to be appreciated—and that means opening up the moment, stretching it by being fully aware within it. This requires not esoteric spiritual practices but a simple decision to do so, followed by action.

It took two centuries for us to arrive at our current frenzied pace. In the Middle Ages, there were no such things as clocks. Even the hourglass wasn't in use until the late thirteenth century. Even today, in areas of Africa and New Guinea, some languages have no equivalent words for hour, minute, or second. In Western culture, the Industrial Revolution fundamentally changed the notions of time. Eventually workers began to be paid by the hour rather than the day, week, or month, and their value was determined by how much they could produce in that hour. Time became money, literally. People stopped speaking of "passing time" and began speaking of "spending time." As the acceleration of technological progress increasingly foreshortened time, technology's promise of reducing worker hours never panned out. Economists predicted as recently as the 1970s that the workweek would dramatically decrease. Instead, between 1977 and 1997, the average workweek of salaried Americans increased from forty-three to forty-seven hours—almost an hour a day. The contributing factors in a lengthening workweek include a desire by employers to work a smaller workforce for longer hours to avoid paying benefits and a rising consumerism among workers that has them trading off time for money. As Juliet Schor wrote in 1991 in *The Overworked American:* "We could now reproduce our 1948 standard of living (measured in terms of marketed goods and services) in less than half the time it took in 1948. We actually could have chosen the four-hour day. Or a working year of six months."

Instead, in the 1990s alone, Americans increased hours worked per year by thirty-six hours—virtually an extra week of work. Globalization, with its harsh focus on the bottom line and its threat of cheap labor elsewhere, is another factor pushing U.S. productivity and hours up. In 2000, the average employed American worked three and a half more weeks a year than a Japanese worker, six and a half more weeks than a British worker, and twelve and a half more weeks than a German worker, according to the International Labor Organization. U.S. workers average a paltry thirteen vacation days a year, compared to forty-two for Italy, thirty-seven for France, thirty-five for Germany, and twenty-five for Japan, according to the World Tourism Organization.

Our work days are being further elongated by technology, which is rapidly blurring the line between home and work. Cellular phones, laptops, modems, pagers, faxes, voicemail, and E-mail all tether our minds to the workplace, even when our bodies are elsewhere. They make instant responses not just a possibility, but a demand.

Lost amid the push for efficiency is our effectiveness; quality is lost to quantity. Our tools of daily life are now made with less quality, our interactions on the phone and in person, and now E-mail, are quick and little valued.

Small wonder that with our relaxation time, we seek escape by rushing to the mall and buying luxury items that make us feel good about ourselves. As Schor pointed out in 1998 in *The Overspent American,* when husbands and wives work long hours, the Joneses that they want to keep up with are no longer neighbors, whom they barely know, but coworkers higher up the corporate ladder and the sitcom actors inhabiting luxurious stage sets. Credit card debt per U.S. household was $8,488 in mid-2001, versus $5,832 in 1995—indebtedness that adds pressure to work even more hours.

To escape stress, we also turn to fast-tempo activities that forcefully grab our attention: Stock-car racing, gambling, video games, channel surfing. When we want music, we turn to the machine-gun tempos of rap, techno, hard rock. The renowned anthropologist Edward Hall noted that one can understand the internal rhythm of any society by listening to its music, and we no longer live in a time of Gregorian chant or Beethoven. Instead, our music, like our society, draws on a narrow range of accelerated rhythms. It reflects our inner hurry—and choreographs it.

There is a word for the process by which rhythms fall in synchronization with one another: entrainment. It is one of the great organizing principles of the world, as inescapable as gravity. Entrainment explains how one rhythm works with another, and how separate entities, from molecules to stars, fall into rhythm.

Our entrainment—our coming into sync with another person, object, sound, mood, rhythm—can be short term or long. It can take the form of a shared smile, a dance, an act of love, an intense discussion, teamwork, a feeling of community. Frederick Erickson of the Interaction Laboratory at the University of Pennsylvania has shown that entrainment takes place even at the dinner table. When family members talk, the syllables they stress carry the same rhythm. When the conversation lapses, the shared rhythm continues: Someone reaches for the salt on a beat; a knife hits a plate on a beat; and, when the meal is over, the family members' departing footsteps continue to tap out the beat.

Entrainment is something that's so much a part of our lives that we don't usually notice it. Yet we march to its rhythms—and today, that rhythm goes snap-snap-snap all the time. Unconsciously, like a poison ingested by our bodies in a deceptively sweet syrup, we have adapted to a faster rhythm. It controls the way we walk, the way we speak, the way we respond to intimates and strangers, the way we don't relax. It habituates us to simply skim the surface of the experience and then to move on. For example, a recent study of zoo-goers at the National Zoo in Washington, D.C., found that the average time people spend looking at any one exhibit of animals is five to ten seconds. You might as well be flipping through a picture book for all the experience that can give you on how an animal moves, eats, communicates.

It's not that going fast is necessarily wrong. Our technology and modern advances hold great promise for us, though only if we can find the balance. Author Rene Daumal writes, "A knife is neither right nor wrong, but he who holds it by the blade is surely in error." Learning to shift our rhythms is about living wisely with time. The path into the future requires personal and systemic change.

THE PERSONAL DIMENSION

Mental Time vs. Emotional Time

We have become habituated to society's rapid pulse, but it doesn't have to be so. By becoming consciously aware of these rhythms in ourselves and the people around us, we can shift those rhythms and, therefore, shift time. But we must slow down in order to listen and feel. Understanding is impossible without serenity; serenity only exists when time moves slowly.

And that, for many people, is where the problem lies. When we slow down and take a deep breath, what often rises to meet us are our feelings. And that's often the last place we want to be. Essentially, we have a deal about the present moment. "I'll be here if it's good." The problem is that if this moment isn't good, is painful or upsetting or not up to our expectations—then we don't want to be here—we resist life as it is right now. Much of the stress in modern life is simply our resistance to what's so in this moment. Life contains good and bad experiences for all of us, no matter how much we try to control circumstances and events, so much is ultimately beyond our control. Success in life isn't about having it all worked out, rather, it's often about the ability to accept this moment just the way it is.

Our thoughts and feelings operate at drastically different rates and carry very different rhythms. In "mental time," the brain communicates electrically, its synapses registering thoughts and ideas in tiny fractions of a second. Try this: Think of a red balloon. A pink elephant. What time it is. It is so easy to go from one thought to another that we rarely even think about how quickly we're doing it.

Now try this: Feel sad. Feel angry. Feel rapturously in love.

Hard, isn't it? These feelings can't be conjured up just like that.

Feelings are experienced by way of chemical communication within the body. They are a hormonal surge, a wave that washes over us. It takes "emotional time" for them to emerge. And to adequately deal with real feelings takes more time—so, when we are rushed, it's much easier to habitually go to our mind and repress our feelings.

The mind, with its lightning-quick synapses, seems to get the job done. Feelings just get in the way—and given full rein, we fear, they might pull us under and drown us. So when we pause and unpleasant feelings inevitably bob up, we bolt from them—by turning on the TV, eating sugar, making a phone call. Anything to not be in the moment.

The danger to this, of course, is that feelings do have their way. Ignored, they shape our unconscious behavior, condemning us to live in self-destructive patterns of the past. Or they erupt from us: Against our better judgment, we blow up or lash out in an experience that Daniel Goleman, author of *Emotional Intelligence*, calls an "emotional hijacking."

Saddest of all, when we don't allow ourselves to feel, we deprive ourselves of the richness and fullness of life. If we don't continually repress our emotions, then ordinary moments take on a deep richness and flavor. Being emotionally present with your child, you can receive from him as much as you give him. Being emotionally present with your spouse, the understanding is so deep you hardly need words to communicate.

Being open to and accepting of our emotions allows us to sit quietly in the present. And then we experience something remarkable that is key to living at ease with time: In the present moment, there is no stress.

Stress comes from resisting what is actually happening in the moment—and what's usually happening is emotion, or feeling. Our continued effort to change what is so in this moment is, in fact, the very cause of the stress we wish to avoid. Pain, either emotional or physical, may be present right now, however, it's the resistance to it that causes stress, while acceptance leads to relief. If, for example, you're going through a divorce, a job loss, a painful illness, problems with children, etc., and you don't allow yourself to feel the pain, then the suppressed pain becomes a lens through which you see all of life. And life seen like that holds little but stress. But when we accept what is so right now, even if we're tired or frightened or hurt, we don't also feel stressed. We may not be happy, but we're open to the reality of life at this moment in time.

By not fighting the moment, we're not adding to our physical trials. In the early stages, stress weakens our adrenal glands, stomach lining, and immune system. Unrelieved, it eventually leads to the breakdown of vital body systems, causing hypertension, heart attacks, strokes, degenerative diseases, and even cancer. We cannot always control the circumstances and events of life, yet we can avoid most of the stress of it all.

Mindfulness and the Moment

There's a story that beautifully illustrates accepting the moment. A Zen monk being pursued by a ravenous tiger climbed halfway down a cliff—and hung by a branch above a ledge he found inhabited by an equally ferocious lion. Growing next to the branch was a bush with a single strawberry, which the monk picked. The monk smelled the strawberry, felt the strawberry, bit the strawberry . . . and thought to himself, "How delicious!"

This parable is not about blindness and denial: The monk is well aware of his predicament. Precisely because of his situation, he is more able to savor the fullness of the moment. Why miss the strawberry?

Many of us believe that when the tiger runs away—when things calm down—then we'll eat the strawberry. When the next crisis ebbs, we'll take time to relax, enjoy life, start the diet, exercise, take our dream vacation. But when the tiger disappears, a lion may take its place. So now, right now, is the time to pay attention to the moment.

There's a word for a conscious focus in the present moment: "mindfulness." It is drawn from Buddhist meditation practices. At first the word sounds confusing: It does not refer to living only in the mind or analyzing everything, or staying in mental time. It means conscious awareness of the present, using all our faculties, all our senses—being aware of what's going on around us and within us as well. Mindfulness is a state of being that we can experience at any moment of our life. Attention and awareness are all that is required.

Each moment has a rhythm: The question is, how wide is our now?

When we enter a state of mindful attention, the present moment, the now, eases open. And when it does, life pours in.

The more we are mindful in simple, everyday situations, the better we get at it. It is easy enough in moments of great crisis or great joy—a car crash that catapults us into a state of full attention, or a wonderfully intimate evening of lovemaking. But it is just as important in everyday tasks—being consciously awake and alive while sweeping the floor, driving a car, or walking down a street is to expand the ordinary moment and make life more full.

One of the simplest and most effective ways to become mindful is to pay attention to our breathing. It brings us mentally and emotionally back to center, as though we were resetting our internal clock mechanism.

Try it now. Become aware of your breath. Really experience it. Fill your lungs completely, slowly, and let the air out slowly, deliberately. Do it a few times. "Watch" your breath as it comes in, watch it as it goes out.

Typically, when you begin such an exercise, you'll find that your mind races—"itching, twitching, and bitching," as one mindfulness teacher put it. But just as with an emotional wave, any discomfort will soon recede. If thoughts race through your mind, just notice them and breathe them gently away on your next exhalation. You'll find the more you do this, the quieter your mind will become.

Another effective way to deepen the sense of the present is through one of the many forms of meditation. As author Stephen Levine notes, "Meditation is a means to an endlessness." Doing it on a daily basis, even for five minutes, is a portal into a greater spaciousness or being more at ease in everyday life. At its heart, meditation teaches one to become so fully immersed in the present that nothing else exists. Developing the ability to focus more fully, to become ever-more present, is the path to less stress and greater contentment.

Another way to expand the moment is by listening to slow, beautiful music, to listen to the tone of a bell, to smell and touch something beautiful. Or to closely watch a small child totally focused on what she's doing, whether it's blowing bubbles or splashing in rain puddles. That was us in our childhood—and can be us in the moment when, like that child, we drop our concern for the future or regrets for the past.

These practices of entering fully into the moment lie at the heart of changing our relationship with time. A two-step process takes us there. First, we become aware of the moment—which means consciously putting aside distractions to grasp clearly the dynamics of the present. Using our bicycle metaphor, it's becoming aware of our speed and the terrain we're flying over. Second, we sense the particular rhythm and flow of the present moment. We shift downward or upward to align ourselves with what is needed.

We can either match our internal rhythm to the external rhythm of the moment or tune into our own rhythm and choose to stay with it. If a moment demands an urgent response, we respond urgently, without resisting or wishing to go more slowly. If the next moment involves being with someone in distress, we can slow down and listen closely. In a particular situation we may indeed choose to go faster, and that is not wrong. As we learn to slow down, we actually get better at going faster. Even when I'm going fast, like when I'm windsurfing, I'm at my best in that moment because I'm relaxed and fully present. Michael Jordan epitomizes how this works when he is dribbling the ball down the floor the last eight seconds of play. It seems like eight minutes, and the key is that he looks like he's on a Sunday stroll. He doesn't want to be anywhere else: It's not like he has to get the game over because he has a dinner date or that he resents being given the ball again. He is just present and alive—and the power of his presence in the now entrains everyone who is watching him to his rhythm.

Living in multiple rhythms is much healthier than life lived at one speed. Just as cross training is associated with a higher degree of athletic fitness, homeodynamism rather than homeostasis, in which rhythmic states go through variability, is associated with greater immune system and overall health.

Learning How to Be at Ease with Time

An effective way to train oneself in timeshifting is to develop rituals that shift us from one rhythm to another. In modern life there are innumerable opportunities for us to develop these rituals. Here are six exercises you might consider trying:

Be in the moment. Find quick, easy ways to break a rhythm and enter into a deeper one. Thich Nhat Hanh says to let the telephone be a bell of awakening. When it rings, stop and take a deep breath instead of snatching it off the hook. You'll find yourself slowing down, becoming calmer and better able to respond to the call.

Other examples: In a heated meeting, take a bathroom break, to breathe more deeply. Take a relaxation pause at your desk when you boot up the computer. Before agreeing to any action, count to ten.

Create time boundaries. Each of us needs some time that is strictly and entirely our own, and we should experience it daily. Preferably, it should be the same time every day—a half hour after dinner, perhaps; fifteen minutes before the start of work; an hour in the afternoon. Make this a time for meditation, for contemplation, for enjoyment of the things around you. Make it "non-productive."

For example, in the morning before work, walk your dog, putt about the house, meditate, listen to music, read poetry—or just sit on your bed and do nothing. In the afternoon, take a walk in the park, go for a swim, ride a bike, drink a leisurely cup of tea. Just make sure the time is yours alone and that you don't interrupt it with your to-do list, or anyone else's.

Honor the mundane. We often treat our days as composed of highlights—the big meeting, a good meal, an outing with the kids—with the in-between times merely connecting the "important" events. These mundane times, however, offer treasures for the soul.

I can sweep the floor to clean it, or I can sweep it to sweep it. The second way, I actually experience the moment. Focus on the task at hand, and see where it takes you. As Thich Nhat Hanh put it:

> If I am incapable of washing dishes joyfully, if I want to finish them quickly so I can go and have dessert, I will be equally incapable of

enjoying my dessert. With the fork in my hand, I will be thinking about what to do next, and the texture and flavor of the dessert, together with the pleasure of eating it, will be lost. I will always be dragged into the future, never able to live in the present moment.

Create spontaneous time. Remember snow days as a child and what a joy they were? As adults, we need to create our own snow days—a time for unplanned, unexpected events. Actually schedule time to be spontaneous.

Pick a Wednesday afternoon three weeks from now, write your own name into your appointment book, and leave work at one P.M. for an unplanned afternoon, going wherever your whim takes you. Or pick a Saturday, get in your car, and drive or get on a train or just leave the house and start walking . . . anywhere.

Create time retreats. Once a year or so, spend a week or more doing something out of the ordinary, something that allows you to shift into a slower rhythm. It might mean going into nature or to the sea, into the mountains or the woods. Beware of "vacations" that have you racing through Disney World or European cities. Choose consciously to go somewhere and be still, and watch time open up to you. Learning to slow down requires allowing oneself to simply be present without the need for anything to be happening.

CHANGING THE SYSTEM

All of us can experience time in new and rich ways simply by learning to be present in the moment. However, there are also government and corporate practices and policies that, if implemented, would make life far more balanced for all Americans. Ultimately, a sustainable system must create structures that support rest, renewal, material security, and flexible work options. Some of these system changes include:

Four-day workweek and flexible work options. Since the late 1960s, the annual working hours of the average U.S. employee have increased by about 180 hours per year. Working hours for parents have increased even more. For example, since 1979, parents in married-couple families aged twenty-five to fifty-four have increased their hours by 354. In response to the time pressures facing their employees, many corporations and private institutions are experimenting with offering options such as a four-day workweek, telecommuting, job-sharing, regular sabbaticals, and other programs that recognize our need for reduced work hours. A four-day week or a family friendly schedule that would allow employees to work nine to three-thirty for five days, getting home in time to meet schoolchildren, would greatly enhance family well-being. These policies should be encouraged, through government subsidies and policies. In the Netherlands, for example, the government itself took the lead by hiring new employees on four-day workweeks. And whole industries shifted over to the four-day week. Other European countries, such as France, have mandated shorter weekly hours.

Increased vacation time. The average American worker gets less than two weeks of paid vacation annually. This contrasts with most Europeans, who receive four to eight weeks of paid vacation. In some European countries these long vacations are the rights of all workers; in others, vacations are mainly covered by strong union agreements that guarantee paid time off. Millions of Americans get no paid vacation and take no unpaid break because they can't afford to. The U.S. government could enact a statutory right to two weeks paid vacation for all employees and give tax breaks to companies that offered more.

State and county policies that limit homework for public school students. Children as well as adults are suffering from problems of time stress. They now have more homework than they did just a decade ago and the quantity of homework seems to be rising, as pressures for enhanced academic performance intensify. Many school systems are beginning to challenge the time squeeze being imposed on children by limiting homework and encouraging homework-free vacation periods.

Raised minimum wage. Low wages are one of the reasons many families must work long hours or hold multiple jobs. The value of the minimum wage has eroded more than 20 percent in the last two decades and has contributed to the decline of all workers' wages at the low end of the labor market. The working hours of low-income families have risen to compensate. For a variety of reasons (more special-needs children, less money to buy care, less job flexibility), the time pressures of low-income employees trying to combine work and family are particularly stressful. A higher minimum wage, while not a full answer, would help these households.

Leisure instead of stuff. Americans are extraordinarily productive. Yet we've chosen to take our productivity primarily in the form of material goods rather than in the form of leisure time. We have bigger homes, cars, televisions, bellies, and credit lines yet we have less time to enjoy life. We need to establish a culture that celebrates renewal, relaxation, and leisure, one that supports those who need to slow down and work less. More stuff requires more money, which requires more work and the cycle goes on. One key step into a sustainable future is to reclaim balanced living and simple pleasures.

TIME AND SOCIETY

Developing the ability to choose our own rhythms is not just about living a fuller, richer life. It's about creating a better, more alive, more caring society.

The issue of time is critical in understanding a society in which life is out of balance. Our culture has simply assumed that faster is better. We never seem to question whether the capacity to do more in less time is good. Progress demands producing more at an ever-quickening pace. Yet in the midst of this speed we have lost something vital in the fulfillment and enjoyment of life itself.

When we are hurtling at breakneck speed, whether it's driving to work or racing through an airport, it's hard to see other people as anything but obstacles in our path. When we're having problems meeting a deadline, we have no patience for a child's tears, a spouse's epiphany. When we're frazzled, it's hard to see the light in anyone's eyes. Multiply this times hundreds of millions and it's not hard to see why discourtesy, alienation, fragmentation, and even violence are mounting in our society.

Time is like wool, Rabbi Zalman Schachter-Shalomi observes. It comes in long strands—yet we cut it into short pieces. When the skeins were longer, they bound families, communities, societies, and nations together. As many as three generations lived harmoniously in one home, bound by religion and shared interests into a united whole. Different groups—religious, social, racial—submerged their differences to work together for the common good of the community and country.

Little binds us anymore. Relationships are transient, divorce always an option. Our children and parents move far away, groups fight each other, and the term "community" no longer has the significance that previous generations of Americans experienced living together. Job security seldom exists; nor does company loyalty.

Short-term thinking rules the day. In England, the huge oak timbers used as beams in an Oxford College great hall built five hundred years ago were rotting. The administrator found in old records that an oak grove had been planted near the college at that time—specifically so the large timbers would be available when needed, centuries in the future. Today, in America, our house sidings last fifteen years, our computers two. Sports arenas erected thirty years ago are deemed too old because they lack domes, luxury boxes, or sufficient advertising space.

I believe that the speedup of time is intricately connected with the holes in the ozone layer, undisposable nuclear waste, endangered species, polluted rivers. The more we operate only from our minds, the more we divorce ourselves from our emotions and our senses, the more we will bulldoze, blacktop, pollute, and mindlessly destroy life. Land, nature, will increasingly be no longer a deeply felt experience but a mere concept, even a profit center. It is only by slowing down to appreciate our environment and to literally smell the roses that we will be saved.

Yet the political response is to go full speed ahead even faster—for reasons also connected to the speedup of time. The democratic process takes thought, discussion, argument, synthesis—time. To truly understand issues, we must hear both sides, study facts, weigh, ruminate. To evaluate a candidate we must listen to the ideas, ask probing questions, consider the person's truthfulness. Instead, today we have sound bites.

When a culture lives only in short moments, it's easy to understand why violence escalates, why movies depict ever more graphically sex and mayhem, why boom boxes and ambulance sirens get louder. Something has to penetrate to get attention in a world without feeling. Violence flourishes because it is a way to feel. Consumed by the need for a quick fix, needing the addictive high that comes with risk, more and more, people are turning to fists and bats and knives and guns.

In the age of speed, violence begets violence. Unless we slow down enough to realize what is truly happening in our world, we will continually repeat the mistakes of the past without taking the opportunity to learn and live the blessings that have been bestowed upon all of us. Until we are able to forgive each other and ourselves—which means slowing down long enough to deal with our regrets and understand our grudges—we are perpetually doomed to a cycle without end.

The apocalyptic worldview as represented in almost all movies that attempt to peer into the future, from *Blade Runner* to *Road Warrior* to *Strange Days*, shows a rapid unraveling of the strands that hold together the human family. The sheer magnitude of the speedup all around us will surely plunge us into a darkness that will make the planet less habitable for future generations, for we will not be able physiologically to evolve to match the speed of change. This could leave us with little hope as a society.

Yet an alternative is possible if we are able to shift, to consciously choose our future. An alternative is possible if we, the human race, are willing to consciously embrace our potential as creatures who feel and care for each other and are willing to slow down enough to let this happen.

We must begin with ourselves, our family, our children. We lose civilization by not truly being civil and kind with each other. And how large is the circle we call family? If it's just our household and everyone else is outside that circle, we're perpetuating isolation and separation. A simple way to expand that circle, and our hearts, is to "practice random kindness and senseless acts of beauty," as the bumper sticker puts it. Being kind, generous, and available to strangers who need a smile, a hand, a tip inevitably shifts us into a deeper mode of consciousness.

If we are willing, there is a kind and beautiful future that we can pass on to future generations. It is a society of plenty—measured not by goods but by quality of life. My view of our positive future looks like this:

- Shorter work hours and consequently fuller employment, based on the European movement that has already begun.

- The resulting growth of a leisure-time industry devoted not to consumerism but to service. More adult education classes, with time to immerse oneself in inner-directed study, art, sports, and nature.

- An education system teaching four R's, not three—the fourth being relaxation.

- Our children being given time as well as space, so they can be children and find their own rhythms.

- People working within their own environments to help slow their rhythms, particularly when working from home with fax, modem, and E-mail.

- Computers programmed to entrain with our slower rhythm, so they become servants in our time.

- A rededication to communities, such as retirement communities or the co-housing projects pioneered in Europe, in which people share services, goals, and ideals. The trend will be toward developing interdependent living situations resembling the villages of old.

- More parks, marshland, wilderness areas, in direct contravention of the trend today, as more of us entrain with the healing rhythms of nature and the earth.

- More and more people involved in contemplative practice and meditation.

- People using holidays as holidays, not as excuses for shopping sprees. Imagine a Christmas season devoted to prayer, a Presidents' Day devoted to the study of history.

- More courtesy and more kindness toward strangers.

- The creation of more rituals for sacred time.

- The honoring of doing nothing, praise for inactivity, cheers for those whose goal is expanding time, not income.

- More people inhabiting a present so deep, so profound, that we feel not only the rhythms of the earth, but can tune in to the rhythm of the cosmos itself, where there is no time, only awareness.

- The ability to shift time, and in so doing . . .

- . . . There is hope . . .

- . . . And love, respect and caring—for humanity, community, friends, family, self, and all of life.

We have the opportunity to live on this planet in a way that all people feel the grace of human existence, instead of racing through life and inflicting suffering upon ourselves and each other. And that can start with something as simple as taking a deep breath, each of us, this very moment.

HOPE IN NUMBERS

Robert Engelman

The immensity of the world's environmental problems often overwhelms us. The world is so big, it seems, and I am so small. How could my behavior harm the planet? But individuals' behaviors *do* matter, because it is all of our behavior collectively that has the impact—each person's use and disposal of natural resources, multiplied by billions of people using and disposing in much the same way, even if often on much different personal scales. It's not just how we consume, but how many consumers we are, that matters. Today, 6.2 billion human beings drink water, eat food, eliminate waste, and need shelter and energy to move, cook, and regulate our body temperatures. More than 75 million more people join us each year. Natural resource consumption derives much of its salience as an environmental issue not *in contrast to* population growth, but *because of* population growth, and not just today's or tomorrow's population growth, but all the growth that has occurred since the human species emerged. The multiplying power of population can transform seemingly benign individual behaviors into forces that undermine both living and nonliving systems and the natural resources people depend on. The global climate, nonhuman species, fresh water supplies, and countless other manifestations of nature simply aren't vast or resilient enough to endure the collective impact of populous and high-consuming modern humanity.

Ultimately, having a world that endures means reducing both the average individual's consumption and aggregate rates of population growth, as well as applying technological change to reduce human impacts on the environment. Having a world that is equitable and peaceful means ending gross disparities in income, power, and consumption itself. These goals require a strategy that still seems a bit novel but is increasingly evident in the post–September 11, post-Taliban era: Those who would change the world for the better must attend to the rights, needs, choices, and capacities of women, especially women in developing countries.

In the mid-1990s, a teenage mother of two in a village in Mali—where the average woman gives birth to six children—told me she was desperate to avoid a third pregnancy, but had no access to any family-planning services. In a neighboring village, other women testified that they, too, wished they could plan their families, "because we are tired of having one in front and one behind"—a local expression for being pregnant while an earlier baby still clings to a mother's back. Survey research suggests that these Malian women are not alone, and that nearly 40 percent of all the world's pregnancies are not intended.

When women manage their own lives, and especially their own childbearing, fewer babies are born and population growth slows. This has been the world's dominant demographic trend for thirty years. Governments don't need to try to convince their citizens to have fewer children; surveys already show that most people in developing countries have larger families than they'd like. The need is for governments to help parents achieve their childbearing desires in good health, and then let population growth trends fall where they may. And fall they will. Today, both the desire and the capacity to have later pregnancies and fewer children are on the rise. The use of contraception has mushroomed from fewer than one in ten sexually active couples in the 1960s to more than half of all such couples today. The world's average family size has collapsed from five children per woman in the 1960s to fewer than three children today. Roughly two children per woman constitutes the "replacement fertility" needed to bring population growth to an eventual end, although that process takes several decades, even after replacement fertility is reached. World population growth rates, which climbed to 2.1 percent annually in the late 1960s, have dropped back to about 1.2 percent annually today (although, with a larger population base, more people are added each year than in the sixties). The populations of Japan and most European countries are peaking now or have begun declining.

Obviously, a lot has changed in the linkage of population, natural resource consumption, and the environment since Paul Ehrlich wrote *The Population Bomb* in the late sixties. Once associated with prospects of environmental catastrophe and collapse, popula-

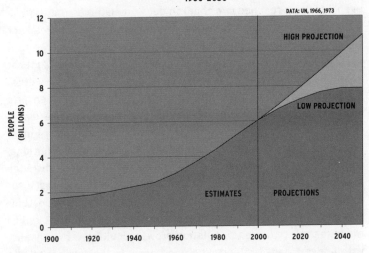

WORLD POPULATION: ESTIMATES AND PROJECTIONS
1900-2050

DATA: UN, 1966, 1973

HIGH PROJECTION

LOW PROJECTION

PEOPLE
(BILLIONS)

ESTIMATES   PROJECTIONS

tion trends have at last reached the point where they are among the few sources of optimism for the planet's environmental future. If governments and other institutions take the right steps in the coming years, many of us could be the first human beings since the Middle Ages to witness an end, or even a reversal, to human population growth. We can be hopeful that humanity will not, indeed, consume the planet to death—at least not through endlessly increasing its numbers.

Among those needed steps is to offer, to all who seek it, access to services that help sexually active people postpone or avoid pregnancy and prevent sexually transmitted infection. The failure of many governments to assure that access is a factor not only drives continued population growth but deepens poverty and illness. Moreover, average world fertility is still significantly above the replacement level. History's largest generation of young people is just reaching childbearing age, without much guidance from their elders about how to avoid unintended pregnancy and sexually transmitted disease, and with many young people unable to gain affordable access to good family planning services. Population momentum—driven by the large proportion of reproductive-age people who are the legacy of past rapid population growth—all but guarantees at least a few decades of increases in human numbers, even if fertility reached two children per woman tomorrow.

Despite this momentum, governments and individuals can act to hasten the end of population growth. Those actions include supporting good family planning services and choices and the rights of young people and women to make decisions about their own lives. Demographic research provides abundant evidence that when girls go to school at least to the high school level, when women are able to seize economic opportunities that are similar to those of men, and when they have access to a range of contraceptive options, they tend to have two children and to bear them between the ages of twenty and forty-five. They have few or none in their teens. If societies could assure that women everywhere had such opportunities and access, and if men everywhere supported women's right to manage their own lives, population growth would slow rapidly and, within a few decades, end. The good news is that almost all the world's governments have agreed to address population issues in just this way. The bad news is that few have gone far enough to make their promises real, and the United States has been among the most recalcitrant of the wealthy countries.

If the global commitment to ensure access to family planning services and to improve the circumstances of women turns into money and action, population trends could become positive environmental forces, supporting rather than undermining the work of modifying individual consumption and reducing poverty in a still-crowded world. For it is increasingly clear, as the story of the Malian women suggests, that the future of population, the environment, and economic development itself is tightly knotted to the lives of women. Not only do educated women with access to contraception have fewer and later childbirths, they are much less likely to be poor and much more likely to be making multiple contributions to their families, their communities, and their nations. Girls attend school rather than stay home to attend to younger siblings. Women start jobs and careers, and then families. Women able to control their own childbearing are more likely to earn and manage their own incomes, and more likely to pay their children's school fees. In short, if you want a more equitable and sustainable society and population in balance with the natural world, attend to the aspirations of the world's women.

A BRIEF HISTORY OF POPULATION

For most of human history, parents had an average of roughly two children who themselves survived to become parents. We know this not by demographic surveys—the first census occurred in the United States in 1790—but by the simple observation that human population grew very slowly until quite recently. The key word here is *survived*. Women undoubtedly had many babies, although some societies employed herbal and other means of contraception. Until recently, however, death rates among infants and children were so high that population growth was at best episodic and localized, rather than consistent and global, as it is with few exceptions today. It was mostly these high death rates among the young that constrained population growth until the modern era. Gradually, improvements in public health behaviors as simple as hand washing and in agricultural technologies as basic as irrigation elevated the prospects of surviving childhood, which boosted population growth.

And so it was that the earth's 1 billion people in 1800 became 1.6 billion in 1900, 2.5 billion by 1950—and 6 billion by 1999. The image of "spiraling" population growth, however, has been inaccurate since the late 1960s, when the global rate of population growth peaked and began to decline. No one can fully explain why this happened when it did. The direct cause, however, was that women began having fewer children than ever before in human history. They were able to do this because modern means of contraception became increasingly available—and increasingly attractive—in most parts of the world.

This demographic revolution, however, is developing unevenly around the world. In Japan and much of Europe, contraceptive use rose so rapidly beginning in the 1970s, backed up by increasingly accessible legal abortion services, that fertility has fallen well below the two-child replacement level. Countries such as Italy, Spain, Armenia, the Ukraine, and Russia have fertility levels so low that many analysts worry about the economic implications, as smaller proportions of working-age people must effectively produce the wealth that supports ever-higher proportions of the old.

A fundamental tension between conventional economic growth and environmental sustainability manifests itself in the dynamics of slower population growth. If economic growth and the social security systems it supports must depend on ever-larger generations of young people, its eventual failure is assured, because the earth and its natural resources are finite. Population growth must end eventually, and the aging of population—that is, a rising proportion of older people—is an inevitable result of slower population growth based on lower birth rates and longer life expectancies. As one journalist has commented, population aging is "a small price to pay for longer lives in a less crowded world." Societies cannot infinitely increase their supplies of renewable fresh water, cropland, clean air, and other forms of natural capital to accommodate ever-growing populations, so they must find ways to adapt to aging and declining populations. The benefit for the environment will be clear, because with smaller populations, achieving sustainability will require proportionally smaller decreases in energy and materials consumption for each person.

POPULATION, CONSUMPTION, AND THE ENVIRONMENT

Interestingly, even at the low levels of human numbers that dominated most of the species' history, population's magnification of personal consumption appears to have led to environmental problems. Scholars of early human history now believe that hunting and habitat destruction were the decisive factors in the great extinctions of large mammals at the close of the Pleistocene Epoch. And none doubt that the pioneer colonists of the Polynesian islands, Madagascar, New Zealand, and Greenland were responsible for the extinction of dozens of species of flightless birds on those islands. Moreover, the great forests that once blanketed the Fertile Crescent, the Mediterranean basin, and Europe itself retreated steadily as ancient civilizations and their growing populations utilized wood products and cleared land for farming.

What was distinctive about human population as it doubled and tripled in recent centuries, however, was that environmental impacts became global in scope and accelerated rapidly. New problems arose: water and air pollution that added to the misery and health threats of urban life, water scarcity that became widespread, collapses of ocean fisheries once thought inexhaustible, and the loading of the atmosphere with gases that trap solar heat and push global climate toward higher temperatures.

To what extent did these developments result from population growth, to what extent from changes in technology, and to what extent from changes in consumption behavior? All three factors are so intricately related that there's no easy way to untangle them. Looking over long time periods, however, there are a few conclusions that can be drawn. Population growth has been a dominant *scale factor* in the impact of humanity on the environment, multiplying the impact of specific individual human behaviors. Scale matters because the environment is so sensitive to it. Activities that would have negligible impacts on an individual level can put natural systems at serious risk at the collective level when the "collective" reaches critical "tipping" points, natural thresholds that are often hard to predict in advance.

Automobile exhaust pipes and factory smokestacks, for example, both polluted the air much more intensively in the 1920s and thirties than they do today. However, the problems of urban air pollution and human-induced climate change only emerged with seriousness in the late twentieth century because population growth and economic growth together multiplied the number of cars and factory smokestacks by several orders of magnitude. Lake Chad is drying up rapidly in northwest Africa, not so much because each individual living around it withdraws more water than previously, but because the number of people with basic needs for food and water has multiplied beyond the lake's capacity to recharge itself.

The relationship between population and renewable fresh water consumption is especially instructive, because water represents a finite natural resource, essential to life, that is becoming critically scarce in many countries. The example of water also teaches us that the role of population growth in the environment often becomes most visible when the need for both sustainability and equity in natural resource consumption enter the equation. In Jordan, Kenya, South Africa, and numerous other developing countries, even low average levels of water consumption are threatening entire societies with water scarcity that shows no signs of abating as long as populations continue to grow.

The need to increase global food production to feed growing populations and combat hunger offers further lessons. Canadian geographer Vaclav Smil conducted a simple inventory of the key crop nutrients—nitrogen, phosphorous, and potassium—needed to feed humanity. Phosphorous and potassium are abundant and long-lasting, Smil noted, but nitrogen must be continuously made accessible to plants by converting the inert atmospheric element to nitrogen compounds that plants can use. Based on his estimate of global need, Smil calculated that there is not enough natural nitrogen from animal manure and nitrogen-fixing plants to do the job. Only the manufacture of artificial nitrogen, achieved through the energy-intensive and polluting Haber-Bosch process, sustains current and future world population growth, Smil concluded.

Whether Smil's inventory and conclusions are accurate or not —and no published research has refuted his analysis—his claim illustrates a basic point about human population and consumption. Human numbers can reach levels that make the goal of truly sustainable consumption challenging at best and impossible at worst. The flip side of this conundrum contributes to a positive vision for the future: Stable or declining population supports and affirms sustainable consumption by opening up more environmental "space" for each person to consume without putting the environment at risk.

More Than North vs. South

Some who follow the connections between population, consumption, and the environment categorize it as a conflict between rapid population growth in the "southern," or developing countries, and wasteful consumption in the "northern," or industrialized countries. The reality is far more complex than this game of mutual blame suggests. Environmental problems everywhere are typically manifestations of human behaviors multiplied by the many people practicing them, with the environment responding based on the laws of nature. Consumption would hardly be much of an environmental issue in the North if the populations of these nations had not themselves grown rapidly for the better part of the last two centuries, to an aggregate of some 1.2 billion people today. The population of the United States is today growing almost as fast, at 1 percent a year, as that of the world as a whole, continually amplifying the influence of its high-consumption economy on the global environment. The 10 percent increase in U.S. greenhouse gas emissions during the 1990s correlated closely with the country's 13 percent increase in population. Per capita emissions, which are roughly five times the global average, declined slightly over the decade. Recognizing population's capacity to multiply environmental impacts hardly implies that the only way to address the oversized U.S. role in greenhouse warming is to slow the country's population growth. The immediate need is for dramatic reductions in per capita greenhouse emissions and similar consumption levels in this country, probably through changes in both technology and behavior. But over the long term, ongoing population growth is not compatible—in the United States or anywhere else—with a stable atmosphere or environment.

In the developing countries of the world's southern hemisphere, both the overall population size of roughly 5 billion people and the likelihood that this will grow by a third or more makes the consumption-environment link in this region dangerously problematic. For those in poverty, access to such basic natural resources as renewable fresh water and cultivable soil is already reaching limits in much of Africa and Asia and in parts of Latin America and the Caribbean. Consumption of water, food, forests, and fish—at subsistence per capita levels multiplied by the more than 1 billion people living in absolute poverty—threatens the very natural resources the poor themselves depend on. In Malawi, for example, the average farm holding in 1987 was barely one hectare, or about 2.5 acres, which is less than the amount needed to feed a family. A decade later, in the absence of significant structural changes in agriculture, further population growth had reduced the average holding to 0.86 hectares.

The role of population growth in reducing the availability of key natural resources is especially evident in the case of renewable fresh water, total supplies of which have been relatively stable on the planet since the last ice age. Based on benchmarks of the minimum amount of water needed per person for economic development to occur, the number of people living in water-scarce or water-stressed countries is in the process of expanding from a few tens of millions in 1950 to somewhere between 2.4 and 3.2 billion people in 2025, depending on the rate of population growth in the next two decades. The declining availability of fresh water is especially alarming for food production, which increasingly relies on irrigation. The availability of the world's forest resources is similarly shrinking, as forests retreat and human numbers expand in almost all developing countries. Already four-fifths of the world's population lacks sufficient access to paper for basic informational, educational, and communication needs, and forest loss is contributing to flooding in vulnerable river plain communities that are most often home to poor and marginalized populations. With seafood traditionally the best source of complete protein for the poor, the world's fish harvest has either leveled off altogether or risen only marginally since the late 1980s, while world population has grown by an additional 1.2 billion people.

The poor are most affected by these trends, while the world's wealthy are disproportionately the agents of contributing factors such as human-induced climate change and the export of tropical woods and other natural resources. Yet environmental degradation is not just a product of extremes of wealth and poverty. The billions of people reaching the middle class in developing countries also contribute to consumption-related environmental problems, such as hazardous levels of airborne lead from automobile exhaust in Mexico City and the growing proportion of global greenhouse gas emissions emanating from cars, factories, and power plants in Asia, Latin America, and parts of Africa.

From the standpoint of the planet as a whole and the rest of life upon it, humans are a single species, with a range of consumption behaviors but continually having more impact as both their economies and their numbers expand. If all the humans who are presently alive consumed at the levels typical of North Americans, the environmental results undoubtedly would be catastrophic. Yet both population and economies will grow beyond today's levels. Reducing the world's appalling poverty levels will require increases in per capita energy and materials use in some places. Therefore, per capita consumption among the wealthy will need to fall proportionally much more to reach sustainable levels. But this shift will be far more achievable as we make progress on two other fronts: One is the development and dissemination of technologies and ways of living that offer people greater health, comfort, security, and freedom, without increasing the unsustainability of their energy and materials consumption. The second is slowing and eventually ending or reversing population growth.

Would success on that second front simply encourage more consumption? Globally it may appear that small families tend to be big spenders and consumers, but the apparent correlation is deceptive. Reducing family size does tend to help reduce poverty, but there's no evidence it actually encourages increased or more wasteful consumption of materials and energy. Moreover, the experience of numerous countries in Asia (Iran, Sri Lanka, Bangladesh, and the so-called "Asian Tigers"), Latin America (Mexico, Colombia, Brazil) and even Africa (South Africa, Tunisia, Mauritius) demonstrates that family size can fall rapidly without stimulating major changes in per capita consumption.

A VISION OF FUTURE POPULATION

How can we succeed in turning population growth around without interfering with people's rights to reproduce as they see fit? It has yet to gain widespread public attention, but the world's governments have been on record for eight years in support of an alternate path that abandons "population control" for what amounts to reproductive liberation—childbearing by choice, not by chance, and new capacities for women to make decisions about their lives, including when and how often to have a child. In a historic 1994 meeting in Cairo known as the International Conference on Population and Development, representatives from the United States and 178 other national governments agreed to a program of action that shifts the focus of population efforts from achieving numerical targets to people themselves. Slowing population growth remains a legitimate interest for governments, according to this approach, but it can only be achieved by meeting *personal* needs for reproductive health services, education and economic opportunities, especially the personal needs of women.

Such an approach cannot come too soon for women like those I met in Mali, but there are already many stories of success in sub-Saharan Africa and elsewhere in the developing world. In Kenya, thirty-two-year-old Pamela Atenya told a reporter in 1998 that she and her husband, a chef, were able to provide for their son, thirteen, and their daughter, nine, despite a doubling or tripling in transportation and food costs over the previous five years. "If I had two more," she said, "I couldn't dress them, couldn't feed them, couldn't buy their school books." The Kenyan government's family planning services had enabled the couple to have the children they felt they could afford.

Imagine a future in which almost every pregnancy is the happy outcome of a conscious decision to bring a new human being into the world. Intentional parenting, if accompanied by good educational and economic opportunities for women as well as men, could yield a world population that peaks at fewer than 8 billion souls around the middle of this century. The world's population then would either level off or gradually diminish for generations to come. Whatever their freely chosen family size and timing, however, parents will cherish each child. Both reproduction and consumption will be more modest in scale, yet also more intentional and appreciated—ultimately allowing a new and enduring harmony among human beings and between humanity and nature. These are broad outlines integrating population and consumption scenarios into a coherent vision that actually could lead to sustainable societies.

Ultimately, on a finite planet, it is in everyone's interests for the pressure of growing population to be eased and eventually to re-treat, as world population gradually falls to levels more in balance with the requirements of natural resources and natural systems. The last and most critical part of this vision of a sustainable future, then, is to imagine what such a world might look like and how to get there.

The demographic scenario that best combines hope and feasibility is the range of low population projections published by the United Nations Population Division and an international think tank called the International Institute for Applied Systems Analysis (IIASA), based in Laxenberg, Austria. Both agencies publish a range of projections of world population over the next century or so, including low, middle, and high projections. Their pictures of the future have been modified in recent years toward a less populous world, as global fertility rates have dropped faster than demographers had previously expected. The United Nations' low population projections envision world population peaking at about 7.9 billion people just before the middle of this century, then beginning a gradual decline (the end of which is not projected). IIASA considers even lower future population levels to be possible; its low projection envisions a peak of less than 7.4 billion people, also around the middle of the century, with population returning to today's levels before the twenty-first century ends. (Under the UN's high projection, population in 2050 is nearly 11 billion and still rising rapidly. Under the IIASA high projection, which continues to 2100, world population in that year reaches 12.2 billion people and continues growing rapidly.) To be fair to both of these low projections, higher death rates from HIV/AIDS and other causes are considered likely factors in the possible peaking of world population in this century. But the overwhelmingly dominant factor is continued declines in birth rates. The world as a whole is assumed to quickly reach and then sink below replacement fertility.

If the future matters, then it would be helpful if its demography resembles those envisioned in these low population projections. To bring such a demographic future about, the world's nations will need to honor the commitment they made in Cairo to refocus population policies and programs around the world on the quality of human life rather than the quantity of human lives—and especially to provide the estimated $17 billion needed annually to make reproductive health services available to all. Developing countries' governments assumed the lion's share of responsibility for this tab. The United States offers around $500 million a year, but as the world's largest economy it should be providing at least three times as much, and it restricts its assistance to groups that promise to offer no abortion counseling or services, limiting the effectiveness of the U.S. contribution.

The concept of reproductive health embraces family planning but also includes safe childbearing for mother and child, the prevention of sexually transmitted disease, and the prevention of unsafe abortion. The Cairo Program of Action also addresses poverty, the half million annual deaths of women for reasons related to childbirth, the needs of youth for sexuality education and reproductive health care, and the HIV/AIDS pandemic. It recognizes that population-growth and contraceptive-use targets and incentives are misguided and ineffective. While governments have a legitimate interest in population trends, individual well-being is the business of family planning providers and their clients. The Cairo document especially embraced the principle that any hope for slowing population growth, stemming HIV/AIDS and maternal deaths, and narrowing the world's income gaps lies in recognizing the rights of women of all ages and helping them to manage their own lives, especially their childbearing lives.

The Program of Action and the conference it articulated mark a historic confluence of demographic research and the commitment of government policymakers, prodded by a movement of mostly women's nongovernmental organizations that may have been at its most vibrant in the mid-1990s. The research indeed confirms that the dramatic expansion of access to family planning services in the last third of the twentieth century explains nearly half of the revolutionary drop in human fertility in that period. Indeed, today's world population is probably smaller by about 700 million people than it would have been without organized family planning programs. And that gap—between what is and what might have been—grows with time.

As we have seen, when girls complete their schooling through at least secondary school they are more likely to have fewer but healthier babies later in their lives. Similar evidence is available for the influence of economic empowerment of women. Based on logic and experience, it is likely that women who are secure in their persons and not vulnerable to gender bias and physical violence are also more likely to make and put into effect choices that result in slower population growth overall. Worldwide, it is political will, changes in law to recognize the rights of women, and well-funded programs that help women manage their childbearing and their lives that are the key to a healthier planet and an early peak to world population.

The end of population growth worldwide will hardly solve environmental problems by itself, and individual levels of consumption are likely to pose environmental, social, and spiritual challenges for a long time to come, regardless of population trends. But any vision of consumption built upon a world-population peak followed by a gradual and temporary population decline is more feasible and appealing than one imagined in the context of population that grows indefinitely. A stable or declining population, for example, would minimize and eventually eliminate the need for continual suburban expansion, as adequate housing could be provided by existing and replacement stock. Road networks could be well maintained without the need for continual widening and expansion. Increasingly efficient consumption of energy, water, food, and raw materials could eventually reach sustainable levels, whose balance would not be continually threatened by increases in scale and expanse.

This is the heart of the positive vision of human population in a world where sustainable consumption at last becomes a reality. We now know that, just as certainly as human population cannot grow forever, we cannot drive population to a level that we seek. We can influence population trends, however, in ways consistent with our highest ideals and values. Governments, health providers, and educators already have helped slow the growth of population through strategies aimed at more immediate individual benefits. These include family planning that helps parents achieve the birth timing and numbers they seek, more girls in school, more economic opportunities for women, and better recognition of women's rights generally. Policies and programs that sustain and accelerate these benefits deserve much more political and public support and much more widespread application. Among the benefits will be an end to population growth in this century. That could mark the beginning of a world where a happy and good life for all is not only achievable, but also capable of enduring indefinitely on a planet that will never grow larger.

FIVE POLICY RECOMMENDATIONS
FOR A SUSTAINABLE ECONOMY

Herman E. Daly

The old American dream was built on a vision of economic growth. For generations, immigrants have poured into the United States seeking to reap the wealth of geographic expansion and booming industry. Now that the United States and the rest of humanity have begun to confront the natural limits to our resources, however, we are in need of a new vision—a vision of a sustainable economy. This is a dream of prosperity that rests on a high quality of life and a well-developed economy, but stability in material throughput rather than growth.

The five following policy suggestions are part of a prescription for moving us in this positive direction. I have listed them in sequence, starting with the least controversial and most concrete, and then moving step by step to the most controversial, and indeed metaphysical. Since even the least controversial proposals have not yet been adopted one might reasonably ask, Why not just focus on, say, the two or three least controversial proposals and forget for now the more controversial ones? Avoid controversy, seek consensus! That would be a reasonable strategy, except for the fact that the three least controversial policies are national policies that would be undercut by global economic integration, and logically require the fourth and more controversial policy of resisting global integration if they are to have a chance of being effective.

The fifth suggestion is a kind of "meta-policy" of facing up to what A. N. Whitehead called the "lurking inconsistency." This radical inconsistency at the basis of modern thought, he said, "accounts for much that is halfhearted and wavering in our civilization." In brief, the inconsistency rests on a widespread modern mindset, especially in economic circles, that while we humans see ourselves as actively shaping our own destinies, scientific materialism dictates that we are incapable of purposefully affecting future events. It greatly weakens any kind of policy agenda by calling into question purpose itself—the belief that humans have any meaningful choices in their actions. Just as the three "least controversial" national policies would be rendered impotent without a fourth international policy that protects them, so all four policies would be rendered meaningless without a worldview that affirms society's capacity to consciously improve itself.

To some it will seem a long stretch from adjusting national economic policy to contemplating metaphysics, but there are definite connections. Many advocates of sustainability, be they ecologists or economists, find their policy wings clipped by adherence to the mechanistic assumptions of their discipline. It is hard to enthusiastically advocate a policy if you harbor even vague doubts about the reality of purpose. It is hard to create a more sustainable society if people believe that economically unsustainable practices are inevitable. However, by acknowledging the power of ideas and purpose to transform the world, our society can affirm its ability to transform unsustainable economic practices.

Let us consider in turn the five policies.

1. STOP COUNTING THE CONSUMPTION OF NATURAL CAPITAL AS INCOME

Our current economic practices assume consumption is always a positive thing—even when it results in depletion of natural capital. Income is by definition the maximum amount that a society can consume this year and still be able to consume the same amount next year. That is, consumption this year, if it is to be called income, must leave intact the capacity to produce and consume the same amount next year. Thus sustainability is built into the very definition of income. But the productive capacity that must be maintained intact has traditionally been thought of as manmade capital only, excluding natural capital. We have habitually counted natural capital—from clean water and soil to forests and the biodiversity of species in the world—as a free good. This might have been justified in a time when human population was limited in relation to abundant natural resources, but today when human population threatens to overwhelm natural resources, it is anti-economic. The error of implicitly counting natural capital consumption as income is customary in three areas: (1) the System of National Accounts; (2) evaluation of projects that deplete natural capital; and (3) international balance of payments accounting.

National accounts, the first area of error, is widely acknowledged and efforts are underway to correct it. The World Bank played a pioneering role in this important initiative, and then seems to have lost interest. Currently, there are several initiatives to reform our national accounts, most notably in the Netherlands, following the pioneering work of Dutch economist Roefie Hueting.

The second, project evaluation by international lending institutions, is well recognized by standard economics, which has long taught the need to count "user cost" (the cost of depleting a natural resource) as part of the opportunity cost of projects that deplete natural capital. World Bank's best practice counts user costs, but average bank practice ignores it. Uncounted user costs show up in inflated net benefits and an overstated rate of return for depleting projects. Therefore, an international development project that degrades society and the environment can appear as "successful" due to this mistaken system of accounting. This biased investments toward projects that deplete natural capital and away from more sustainable projects. Correcting this bias is the logical first step toward a policy of sustainable development. User cost must be counted not only for depletion of nonrenewables, but also for projects that exploit renewable natural systems beyond their capacity to regenerate, or beyond their sink capacity—for example, the atmosphere's ability to absorb CO_2 or the capacity of a river to carry off wastes. It is admittedly difficult to measure the depletion of natural resources but too often, economic development projects opt to ignore the issue completely, again skewing public understanding of the true costs and benefits of new roads, dams, pipelines, and other large-scale projects.

Third, in balance of payments accounting, exports of resources are counted entirely as income, even if they are in fact the depletion of natural capital. For example, the only sustainable yield of a forest is income. Yet the whole forest may be cut down, exported, and counted as income. The part of the forest that was yielding a sustainable cut is itself cut down and exported. There is no subtraction for this reduction of natural capital. This is an accounting error. Some portion of those nonsustainable exports should be treated as loss of a nation's resource capital. If this were properly done, some countries would see that their apparent gains from trade are net losses due to the unsustainable export of their natural capital. Reclassifying transactions in this way would trigger a whole different set of International Monetary Fund (IMF) recommendations and actions. This reform of balance of payments accounting should be the initial focus of the IMF's professed interest in environmentally sustainable development. Instead they seem interested only in pushing for liberalization of the capital account as well as the current account, thereby, as will be argued later, subverting their basic founding charter.

John Ruskin, back in 1862, must have presciently been thinking of modern Gross Domestic Product (GDP) when he wrote, "That which seems to be wealth may in verity be only the gilded index of far-reaching ruin. . . . " Growth in GDP, so-called economic growth, has for some countries literally become uneconomic growth because it increases unmeasured costs faster than it increases measured benefits. In Indonesia during the past twenty years, for example, massive, unsustainable exports of timber resources have contributed to a booming GDP, while the nation's actual economic capital—including natural capital—has diminished as a result. Consequently, many policies justified mainly by their contribution to GDP growth, such as global economic integration, would suddenly lose their rationale if correct accounting were to reveal that growth had begun to degrade the overall economy. Maybe that is why the World Bank lost interest in correcting the national accounts.

2. SHIFT TAXES FROM LABOR AND INCOME TO RESOURCE CONSUMPTION

Currently, most governments put taxes on our income and employment, yet they don't tax the extraction of resources. This sends the wrong signals to everyone in the economic system. In the past it has been customary for governments to subsidize the extraction and consumption of natural resources (resource throughput) to stimulate growth. Thus energy, water, fertilizer, and deforestation are even now frequently subsidized. To its credit, the World Bank has generally opposed these subsidies. But it is necessary to go beyond removal of explicit financial subsidies to the removal of implicit environmental subsidies as well. By "implicit environmental subsidies" I mean the external costs of extracting and consuming these commodities. For example, when a mining company opens up a mining pit and leaves the milling waste for local and federal government agencies to clean up, our government is providing a subsidy in the amount of cleanup costs paid by the public rather than the firm responsible for them.

Economists have long advocated internalizing external costs. A blunt but operational policy for doing this would be simply to shift our tax base away from labor and capital income and on to consumption. The present system of raising public revenue by taxing labor and capital in the face of high unemployment discourages exactly what we want more of. The present signal to firms is to reduce their labor forces and to develop technologies that increase the consumption of energy and materials. It would be better to economize on consumption because of the high external costs of resource depletion and pollution and at the same time to use more labor because of the high social benefits of reducing unemployment. Shifting the tax base onto resource throughput internalizes the costs of natural resource depletion and pollution, raising the prices of those resources and thereby inducing greater efficiency in their use.

Politically, the shift toward ecological taxes could be sold under the banner of revenue neutrality—while "what" is taxed would change, the overall "amount" the public paid in taxes would remain relatively constant. However, some progressive income taxation should be maintained by taxing very high incomes and subsidizing very low incomes. But the bulk of public revenue would be raised from taxes on consumption either at the extraction or pollution stage, though mainly the former. The shift could be carried out gradually by a preannounced schedule to minimize disruption, and correct for any undesirable redistributions.

This tax shift, like sustainable development itself, must first be achieved in the affluent, developed nations of the world, primarily located in the Northern Hemisphere. It is highly unfair to expect sacrifice for sustainability in the South if similar measures have not first been taken in the North. The major weakness in the World Bank's and IMF's ability to foster environmentally sustainable development is that they only have leverage over the South, not the North. While the Nordic countries and the Netherlands have begun to lead the way without being pushed, some way must be found to push the rest of the North also.

While it is true that land and natural resources exist independently of humans, and therefore have no cost of production, it does not follow that no price should be charged for their use. The reason is that there is an opportunity cost involved in using a resource for one purpose rather than another, as a result of scarcity of the resource, even if no one produced it. The opportunity cost is the best forgone alternative use. If a price equal to the value of the opportunity cost is not charged to the user, the result will be inefficient allocation and waste of the resource—low priority uses will be satisfied while high priority uses are not. Efficiency requires only that the price be paid by the user of these "free gifts of nature"—but for efficiency it does not matter to whom the price is paid. For equity it matters a great deal to whom the price is paid, but not for efficiency.

To whom, then, should the price be paid? Since we cannot pay nature directly, we pay the owners of nature. But who is the owner? Ideally ownership of land and resources should be communal, since there is no cost of production to justify individual private ownership "by whomever produced it." Each citizen has as much right to the "free gifts of nature" as any other citizen. By capturing for public revenue the necessary payment to nature, one serves both efficiency and equity. We minimize the need to take away from people by taxing the fruits of their own labor and investment, taxing resource consumption instead. We minimize the ability of a fortunate few private land and resource owners to reap a part of the fruits of the labor and enterprise of others. Land and resource rents (unearned income) are ideal sources of public revenue.

3. MAXIMIZE THE PRODUCTIVITY OF NATURAL CAPITAL IN THE SHORT RUN, AND INVEST IN INCREASING ITS SUPPLY IN THE LONG RUN

Economic logic requires that we behave in these two ways toward the limiting factor—i.e., maximize its productivity and invest in its increase. Those principles are not in dispute. Disagreements do exist about whether natural capital is really the limiting factor. Some argue that manmade capital is a perfectly good substitute for natural capital. In the past, natural capital has been treated as superabundant and priced at zero, so it did not really matter whether it was a complement or a substitute for manmade capital. Now, remaining natural capital appears to be increasingly scarce. For example, the fish catch is limited not by the number of fishing boats, but by the remaining populations of fish in the sea. Cut timber is limited not by the number of sawmills, but by the remaining standing forests. Pumped crude oil is limited not by manmade pumping capacity, but by remaining stocks of petroleum in the ground.

In the short run, raising the price of natural capital by taxing resource throughput, as advocated above, will give the incentive to maximize natural capital productivity. Investing in natural capital over the long run is also needed. But how do we invest in something that by definition we cannot make? If we could make it, it would be manmade capital! For renewable resources, we have the possibility of "fallowing" investments, allowing this year's resource flow to accumulate rather than consuming it. For nonrenewables, we do not have this option. We can only liquidate them. So the question is, how fast do we liquidate, and how much of the proceeds can we count as income if we invest the rest in the best available renewable substitute? And of course how much of the correctly counted income do we then consume and how much do we invest?

One renewable substitute for natural capital is the mixture of natural and manmade capital represented by plantations, fish farms, etc., which we may call "cultivated natural capital." But even within this important hybrid category we have a complementary combination of natural and manmade capital components—e.g., a plantation forest may use manmade capital to plant trees, control pests, and choose the proper rotation—but the complementary natural capital services of rainfall, sunlight, soil, etc., are still there, and eventually still become limiting. Also, cultivated natural capital usually requires a reduction in biodiversity and often leads to unanticipated environmental damage, such as the pollution caused by farm-raised shrimp and salmon.

For both renewable and nonrenewable resources, national governments and financing institutions need to invest in enhancing the productivity of these resources. Increasing resource productivity is indeed a good substitute for finding more of the resource. But the main point is the most limited resource is the most important to conserve, and to the extent that natural capital has replaced man-made capital as the limiting factor, our investment focus should shift correspondingly. I do not believe that it has. In fact, the World Bank's failure to charge user cost on natural capital depletion, noted earlier, surely biases investment away from projects that replenish natural capital.

4. MOVE AWAY FROM THE IDEOLOGY OF GLOBAL ECONOMIC INTEGRATION BY FREE TRADE, FREE CAPITAL MOBILITY, AND EXPORT-LED GROWTH

This will move us toward a more nationalist orientation that seeks to develop domestic production for internal markets as the first option, having recourse to international trade only when clearly much more efficient. At the present time, global interdependence is celebrated as a self-evident good. The royal road to development, peace, and harmony is thought to be the unrelenting conquest of each nation's market by all other nations. The word "globalist" has politically correct connotations, while the word "nationalist" has come to be pejorative. This is so much the case that it is necessary to remind ourselves that the World Bank and the IMF exist to serve the interests of their members, which are nation states, national communities—*not* individuals, not corporations, not even NGOs. The Bretton Woods institutions—the World Bank and the International Monetary Fund (IMF)—have no charter to serve the one-world-without-borders cosmopolitan vision of global integration—of converting many relatively independent national economies, loosely dependent on international trade, into one tightly integrated world economic network upon which everyone depends for even basic survival. If the World Bank and the IMF are no longer committed to serving the interests of their members, then whose interests are they serving?

Globalization, considered by many to be the inevitable wave of the future, is frequently confused with internationalization, but is in fact something totally different. *Internationalization* refers to the increasing importance of international trade, international relations, treaties, alliances, etc. International, of course, means between or among nations. The basic unit remains the nation, even as relations among nations become increasingly necessary and important. *Globalization* refers to global economic integration of many formerly national economies into one global economy, mainly by free trade and free capital mobility, but also by easy or uncontrolled migration. It is the effective erasure of national boundaries for economic purposes. With the removal of economic barriers, multiple national economies become one global economy. While *international trade* among independent national economies guarantees that each nation gains from voluntary trade, with a globally integrated economy, national borders blur while corporate power escalates, often doing damage to small nations that are forced into a global export/import economy.

In short, globalization is the economic integration of the globe. But what exactly is "integration"? The very word integration derives from "integer," meaning one, complete, or whole. Integration is the act of combining into one whole. Since there can be only one whole, only one unity with reference to which parts are integrated, it follows that global economic integration logically implies national economic disintegration. By disintegration I do not mean that the productive plant of each country is annihilated, but rather that its parts are torn out of their national context (disintegrated), in order to be reintegrated into the new whole, the globalized economy. As the saying goes, to make an omelette you have to break a few eggs. The disintegration of the national egg is necessary to integrate the global omelette.

However, the model of international community upon which

the Bretton Woods institutions rests is that of a "community of communities," an international federation of national communities cooperating to solve global problems. The model is not one of a global corporate economy and culture that stamps out small enterprises, farmers, and national traditions.

To globalize the economy by erasure of national economic boundaries through free trade, free capital mobility, and free, or at least uncontrolled migration, is to wound fatally the major unit of community capable of carrying out any policies for the common good—national governments. Increasingly, these governments are dominated by corporate interests that oppose vital international agreements required to deal with those environmental problems that are irreducibly global (CO_2, ozone depletion). International agreements presuppose the ability of national governments to carry out policies in their support. If nations have no control over their borders they are in a poor position to enforce national laws, including those laws necessary to secure compliance with international treaties.

Globilization not only weakens national governments, often wreaking havoc on indigenous enterprise and culture, it also has extraordinary environmental costs. The huge increase in the flow of goods requires fossil fuel to transport the goods, packaging from forest products to ship the goods, and applications of chemicals to food products to lengthen shelf life. When whole nations convert to one or two export crops, entire ecosystems are often destroyed in favor of monoculture farming, with exports such as coffee and bananas being obvious examples.

Globalism weakens national boundaries and the power of national and subnational communities, while strengthening the relative power of transnational corporations. Since there is no world government capable of regulating global capital in the global interest, and since the desirability and possibility of a world government are both highly doubtful, it will be necessary to make capital less global and more national.

I know that is an unthinkable thought right now, but take it as a prediction—ten years from now the buzz words will be "renationalization of capital" and the "community rooting of capital for the de-

velopment of national and local economies," "minimum residence ✓
times of foreign investments," "Tobin taxes," etc., not the current
slogan of "export-led growth to increase global competitiveness."
"Global competitiveness" often reflects not even a real increase in
resource productivity, but rather a competition to reduce wages,
externalize environmental and social costs, and export natural capital at low prices while calling it income growth. In short, the entire
direction of economic growth on a global scale must be reversed,
placing value on local and national enterprises.

5. FACING THE LURKING INCONSISTENCY

However, to achieve sustainable development—ending poverty
and conserving the biosphere—we must first address a more metaphysical issue. We must rescue the idea of purpose itself from a
lurking inconsistency.

Too many so-called experts in the academy simply dismiss pur-
poseful action as wrong-headed. These mechanistic philosophers,
neo-Darwinist biologists, and even some of my fellow economists
argue that scientific realism will prevail—that all human activity
is essentially determined by physical forces beyond our control.
Oddly, these same champions of deterministic belief frequently advocate very specific policy agendas aimed at benefiting corporate
interests.

It is vital to challenge this peculiar academic disease that sadly
distracts many of our most promising scientists, economists, and
philosophers. Too many apply their analytic capacities to narrow
and arcane research agendas with little if any relevance to the
great human drama of our times—the urgent need to harmonize our
economic rules and practices with a planet that shows signs of
growing distress and collapse. The search for a sustainable future
ultimately depends on the growing chorus of humans—from all sectors—who are vocally reclaiming the highest of human values and
purpose—those of compassion, empathy, sharing, and deep connections between all living things. It is this hunger for and embracing of
purpose, in defiance of determinism, that gives all of us true hope
for a better future.

The term "lurking inconsistency," as well as its meaning, is taken from Alfred North Whitehead (*Science and the Modern World*, 1925), who explains that a dangerous inconsistency results from scientific realism and its deterministic belief that "physical causation is supreme." This calls into question the possibility of human purpose causing any meaningful change in the world. If purpose is not causative then all policy is nonsense, and all theory on which policy is based is useless.

If purpose does not exist then it is hard to imagine how we could experience the lure of value. To have a purpose means to serve an end, and value is imputed to whatever furthers attainment of that end. Alternatively, if there is objective value then surely the attainment of that value should become a purpose. To serve the purpose of conserving the biosphere, we will first have to reclaim purpose itself from that darkness.

Economists, unlike mechanistic biologists, do not usually go to the extreme of denying the existence of purpose. They recognize purpose under the rubric of individual preferences and do not generally consider them to be illusory. However, preferences are thought to be purely subjective, so that one person's preferences are as good as another's. Unlike public facts, private preferences cannot be right or wrong—there is, by assumption, no objective standard of value by which preferences can be judged. Nevertheless, according to economists, preferences are the ultimate standard of value. Witness economists' attempts to value species by asking consumers how much they would be willing to pay to save a threatened species, or how much they would accept in compensation for the species' disappearance. The fact that the two methods of this "contingent valuation" give quite different answers only adds comic relief to the underlying tragedy that is the reduction of value to taste.

Purpose has not been excluded for economics, just reduced to the level of tastes. But even an unexamined and unworthy purpose, such as unconstrained aggregate satisfaction of uninstructed private tastes—GDP growth forever—will dominate the absence of purpose. So, in the public policy forum, economists with their attenuated, subjective concept of purpose (which at least is thought to be causative) will dominate the neodarwinist ecologists who are still crippled by the self-inflicted purpose of proving that they are purposeless.

A growing number of economists are challenging some of the fundamental tenets of neoclassical economic theory, including the limits on human potential imposed by the almost religious belief that all human preferences are equally good and should and will be maximized in the marketplace. This doctrine essentially institutionalizes and justifies greed and furthermore argues that all alternatives to self-interest and greed are scientifically preposterous. On the contrary, the worldwide struggle for common values, for genuine human welfare, for new rules of commerce, and for sustainable development are heartening examples of human purpose in action.

ANOTHER WAY OF BEING HUMAN

Peter Forbes

Speak these American place names to yourself: Androscoggin, Bear Paw, Bitter Root, Black Canyon, Catawba, Blue Ridge, Congaree, Elkhorn Slough, Gathering Waters, Hollow Oak, Jamaica Bay, Keweenah, Lummi Island, Mojave, Otsego, Prickly Pear, Siskiyou, Tecumseh, Wolf River. These places speak of our history. They are the waters, the food, the wood, the dreams, and the memories that literally make up our bodies. They are our alchemy of land, people, and story. When we show our care for these places that define us, we give ourselves the gift of memory and connection. These landscapes give us a hold in the world. These are the places that inspire our belonging, replenish our souls, and remind us that where we live is like no other place in the world.

The terrifying reality is this: As we fail to protect them, every place in our country begins to look like every other place. Sprawl is gobbling up the American countryside at 365 acres per hour. Over three million acres of forests, farms, and wetlands—the places that inspire our remembering—will be paved over this year alone. There are more malls in America today than high schools. This changes who we are. For example, popular lore has it that the average person can recognize one thousand corporate logos but can't recognize ten plants and animals native to his or her local area.

Our relentless desire for things ultimately devours the land while silencing stories that may be important for us to hear. In losing our connection to the land we also lose an important source of information about how we might live differently. We are left with fewer and fewer sources of meaning for ourselves. Increasingly, there is only one story to hear and one story to tell. For too many of us, this has become a world where the point of trees is board feet, the point of farms is money, and the point of people is to be consumers. We learn that the only story that matters is the one playing in our head, that the only land that matters is what we own, that the only person that matters is ourselves, and that the only time that matters is now.

This loss of our bearings comes out of the loss of the little places we knew of as children, the loss of the landscapes that told us we were home, the destruction of a forest, the disappearance of the grizzly bear and the wolf, the vanishing ways of life, the loss of our relationships with the world around us. *The world I knew is gone.* I hold these words as a talisman for our era. In fact, many of the places we once knew are gone, or rapidly disappearing. What we are losing is the world all of us once knew, no matter how old we are, no matter the color of our skin, the amount of money in our pocket, or the place where we live. What we are losing is the deep relationships people have had with land, and through land, with a community and a story much bigger than ourselves.

I have heard others say that we are deep into a Great Forgetting, a time when our relationships are so fractured that we have almost forgotten why they were important in the first place. I prefer to believe that we are on the brink of a Great Remembering, a time when we can reconsider what matters most to us, when we are punished enough by "the world I know is gone" that we can find the courage to ask ourselves how much is enough? We have nearly said the final good-bye, having long since shaken hands, and are just now waiting to shut the door. We are almost ready to accept all of the loss around us as unavoidable, a fact of modern life, a cost of doing business. But not quite. We hear a piece of music and feel a deep unexplained stirring inside of us. We climb a mountain in the fall and are swept away by the beauty. We find ourselves working with a neighbor, allowing our hands and backs to say what our mind cannot: I care about you and this place. Our love for this land, this life, all of life, has us by the throat even when we don't have the words to speak.

The path to healing is through the act of land conservation, an act that in a hundred different ways shows our sense of hope and affinity by reconnecting us with the life around us. It's a rekindling of what is most meaningful inside each of us. Conservation is a way for humans to reengage with the world, a way to extend our best definitions of our humanity.

Our relationship to land is certainly not the only answer to all of our problems. It would be naive to suggest this. But it *must* be among our first responses. Land is the foundation of our cultural house. And place tells us a great deal about what is good and healthy about ourselves. Anyone who doubts that we still get our most fundamental cultural information from the land should drive out to the closest strip mall, stand in front of it, and ask themselves, "What does this place say about me?" Or, as the Amish ask it, "What will changes in this place do to my family?" What we choose to do with our landscape becomes, fundamentally, what we choose to do to ourselves.

At its best, land conservation is a proud form of civil disobedience that quietly but steadfastly opposes the prevailing cultural forces of our times. Conservation is our refusal to accept what is normal today. And it is a humble and joyful offer of an alternative vision for the future. It harnesses the mystery, the unspoken love, that brings people out of their homes to transform a vacant lot into a community garden or to protect the farms, rivers, and mountains of their lives. Such favorite local places don't necessarily contain any threatened species of plant or animal, but their loss would mean an *extinction of human experience.* And people know it.

Glenn and Kathy Davis are learning Hawaiian traditions and rebuilding Hawaiian culture by bringing young Hawaiians back to the taro fields. For thousands of years, up until the 1960s, taro was a primary food on the islands and it was once grown everywhere in irrigated, terraced fields. The culture developed with this plant. The Hawaiian word for "family," *ohana,* is also the word for the taro root. *Quleana* means both "land" and "responsibility." But since the 1960s, many Hawaiians left taro farming and the traditions of their ancestry for jobs in tourism. Conservation has enabled some of them to go forward to the land, returning to the abandoned taro fields and restoring a set of values that strengthens them as a people. Glenn told me, "People are beginning to remember that we are organic and part of the land. It is essential to have places where you can go and remember that you are something. Now that we've really come back and are committed to the taro again, there are more birds singing in the jungle. The water is flowing again. We have come home."

Land conservation helped a community in Colorado Springs to answer a provocative question in our consumer society: How much is enough? A small handful of dedicated people said they cared enough about one piece of land—Stratton Ranch—to leave their private lives behind and serve a larger world. By halting the building of private luxury homes on land they loved and used as a community, they were envisioning a future not in terms of economic growth but in terms of relationships. These neighbors were expressing their allegiance to ideals, to one another, and to their citizenship *in a specific place.* Their process of building rootedness, through conserving the last open land in the midst of private development, proved to them that the enduring wealth of their community would be based on commitment and sense of responsibility toward one another. Richard Skorman, who was elected to the city council after this land conservation project, said, "I had given up on this city, thought we just had caved in to the whole argument that anything valuable and worth doing was just plain impossible. Saving the Stratton land was a big step in saving ourselves. We can do good. It's there every day now, reminding us of our best intentions as a community."

Miguel Chavez, a fifth-generation citizen of Santa Fe, says that his community's past and promise for the future are tied up in a specific piece of earth: a rail yard and its surrounding thirty-four acres. Many Hispanic, Native American, and Anglo people are not happy with the rush of commercialization and growth that threaten the soul of their community. The struggle over the future development of this rail yard—upscale stores and resort hotels or farmers' markets and open land—has come to be a struggle over the soul of Santa Fe. As one leader said, "This land brings us together and pushes us apart more than anything else could. The rail yard symbolizes the contradictions that we live with. How we handle that contradiction and how we use this land is undeniably the story of how we deal with ourselves." The process of learning, planning, and arguing together is the city's full expression of trying to reform itself. Miguel Chavez says that "without a sense of place, people become criminal to one another. We lose a sense of loyalty to one another. Our ethics arise from a sense of belonging here. Our ethics will arise from this exact piece of land. This is our opportunity to prove that Santa Fe is not just about tourists and making money and forgetting the past, but about a way of living together." The process of protecting this land, learning its symbolism, arguing over its future is, itself, the process of social change.

These examples point the way toward a new radical center in the work of land conservation, where the value of protecting the tiniest urban lot and the largest tract of wilderness are viewed together through the lens of how well they build relationships—relationships between people, between species, between the whole of the land community. This isn't saving land *without* people, or saving land *for* people. The radical center calls for saving land *and* people. We might call this "land and people conservation." It gives us hope of finding a way for all of us to be at home on the land no matter where we live or how long we've lived there. It gives us a powerful new benchmark for land conservation, where the alchemy of human cooperation, activism, and the wild leads people to dwell and imagine differently, to find their own souls.

Aldo Leopold envisioned all of this. Writing more than fifty years ago, he said, "Conservation is one of the squirmings which foreshadow this act of self-liberation." Let us begin to think of land conservation, then, truly as an act of self-liberation. Self-liberation from ways of living that deny us meaning, connection, and purpose. The work of saving land also takes us beyond ourselves. It demonstrates that we value diversity, that we can show restraint, that we seek to engage with the world around us on peaceful terms, that we have many definitions of our humanity, that our values as a people include sacrifice, humility, respect, joy, and love. Conservation is the tangible, highly visible expression of our own *ethics of enough*. Land conservation is helping communities to address racial tensions, to plan more effectively, to learn more about one another, and to strengthen local economies. The struggle over land enables people to tackle other struggles in their lives. I call this the transformative power of land conservation: showing us how protecting what we love can change who we are and how we live.

Classie Parker is at the center of such a transformation. She lives in Harlem just a few blocks from the hospital where she was born. For many years, Classie felt stuck on a street where no one knew anyone else and drug dealers ran everything. She especially feared for her father, who was growing old and needed a way to stay active and get outside. She worried that he would die alone in a building where nobody noticed. In 1992, Classie's apartment stood adjacent to a 3,600-square-foot vacant lot that was crowded with crack vials, needles, abandoned cars, and garbage of every kind. When Classie got the idea to create a garden on that lot for her father to work in, she recruited her brother and a Hispanic couple who lived nearby and their five children to help her. Classie had a vision for a place where the old and young could work together. Today, the thriving garden there is called "Five Star," in honor of the five adults and five children who started it.

One very hot Saturday in July, I found myself on 121st Street in central Harlem trying to get perspective. For an hour or more I sat on the corner of Frederick Douglass Avenue eating peaches and taking in the neighborhood. Everyone and everything was in constant motion: motorcycles racing each other down the avenue, vendors selling sunglasses and old record albums, children playing games at my feet, endless flows of people. But amid all the noise and pavement, there was a quiet green garden. An eight-foot-high chain-link fence could barely keep the sunflowers from pouring out onto 121st Street. With two large town houses protecting either flank, the garden itself was just plain bold and beautiful. A dozen discarded lawn chairs had been retrieved and organized loosely around leaning tables and empty crates as if a card game or a good meal had just been finished. I could see rows of corn, plots of vegetables, climbing snap peas, grapevines, fruit trees, and a dogwood. I could hear birds. Men and women of all ages were hanging on the chain-link fence talking to friends on the street, and then turning back into the garden with a hoe or a laugh.

Five Star is breathtakingly beautiful and heavy with life. It is stewardship and wildness wrapped together and dropped down on 121st Street. Classie produces food, beauty, tolerance, neighborliness, and a relationship to land for people throughout her part of

Harlem, all on less than one-quarter of an acre. Five Star Garden is almost absurdly small, but for the people of 121st Street—who, for the most part, never leave Harlem—the garden is their own piece of land, to which they have developed a very deep personal attachment. As Classie says, "There's love here. People gonna go where they feel the flow of love. There is a difference. You come in here and sit down, don't you feel comfortable with us? Don't you feel you're free to be you? That we're not going to judge you because you're a different color or because you're a male? Do you feel happy here? Do you feel intimidated? Don't you feel like my dad's your dad?"

What can conservationists learn from Classie Parker and Five Star Garden? At my organization—The Trust for Public Land—we ask ourselves what connects our acts of conservation at a place like Five Star Garden in central Harlem with a landscape like the Columbia River Gorge in Oregon and Washington. The gorge is a large, majestic ecosystem, home to a great diversity of species, and it offers humans a physical experience, a potent reminder of the inspirational power of wildness to nurture the body and soul. If we define conservation success in terms of acres, the distinctions between Five Star Garden and the Columbia River Gorge are obvious and enormous. If we are committed solely to biological diversity, the gorge has something very important to offer while Five Star Garden does not. If we enjoy the debates between wilderness and stewardship, there's an argument for both the gorge and Five Star Garden. And if we are interested in sociology, Five Star Garden is a compelling example of a human relationship to land that is transforming a community and greening a small part of the earth. And yet all these explanations and ways of defining success are hugely inadequate. None examines the whole of life, and therefore each perpetuates an unstable system. We need instead to find that radical center that offers a new aspiration for land conservation. The new guiding principle is relationships.

The enduring value of the relationship to the land might best be measured by the extent to which it evolves beyond self-interest. All healthy relationships entail sacrifice and are never solely about what makes one person feel good, but are about what's also good for someone else. Relationship implies a responsibility that goes beyond one's own dreams. Wendell Berry put it this way: "To grow up is to go beyond our inborn selfishness and arrogance; to be grown up is to know that the self is not a place to live."

Viewing land as the place of relationships requires that conservationists firmly put their work in a context of time and history. What we are "saving" is not so much the piece of land but the quality and integrity of the relationship to the land so that what we will and *will not do* is preserved in perpetuity. If we're lucky, the land will evolve and change forever, but it's our human attitude—our values—that most need to be "protected." Our laws protect land *from* us when we are at our worst rather than keep us together when we are at our best.

The quality and character of the relationships that we can have are as diverse as the land itself. For example, our relationship with wilderness is one of forbearance and reverence; our relationship to a working ranch speaks of our commitment and patience. Our strengths as a people will emerge from the quality of our relationships with the land, including our sense of care, well-being, neighborliness, trustworthiness, and health. And it will be equally clear how many of our weaknesses as a culture grow from our inability to develop a connection with a place. Through the lens of relationships, land conservation would have as part of its mission the notions of sympathy, building health and well-being, and reducing suffering much more than the goal of keeping land unchanged.

There is no single science or philosophy to help us protect the relationships between people and the land. Unlike conservation biology, this work of protecting ways of life, or habitats for people, has no highly defined project selection criteria. But it includes growing healthy food, having safe parks and clean rivers accessible to people, building relationships with the land that inspire our sense of ethics and art, maintaining a culture of mutual aid and an appreciation of local beauty, defining our limits as responsible creatures, protecting our cultural and ethnic diversity—all of which contribute immeasurably to the health and well-being of all species on this planet.

This aspiration for land conservation will be reached not solely by how much nature we can put aside, but by how much love and re-

ANOTHER WAY OF BEING HUMAN

spect for the land we can engender in the greatest number of people. Our greatest achievement is not being able to say, "We saved
this place," but being able to say, instead, "You belong here, you are
home." Land conservation can become the story of how the soul of
the land is the soul of our culture, signaling over and over our place
in the world.

HOW DO WE GET FROM HERE TO THERE?

Betsy Taylor

The visionary essays in this book are both practical and profound. They offer us choices. When we hear so much bad news about
the environment and human suffering, a bleak future can seem inevitable. It's not. When we see so much dominating power at work in
our society we can feel helpless. We're not. There are solutions that
will see us through to a safer, more ecologically and socially harmo-
nious future. The challenge is primarily political. The dominant
forces and leaders in our society, along with many average citizens,
are resisting the necessary changes, for a myriad of reasons. This
chapter suggests strategies for overcoming these obstacles.

The quest for sustainable development is the supreme challenge facing humanity today. In 1987, the Bruntland Commission
defined sustainable development as "development that meets the
needs of the present without compromising the ability of future
generations to meet their own needs." Right now, we are not remotely close to achieving this goal and the longer we put off facing
facts, the worse things get. By 2050, the planet will have some 10
billion humans, up from just over 6 billion at the start of this century. Global consumption of water, energy, food, fish, and wood
keeps rising, even as the resources grow scarce and our waste products, especially global-warming gases, skyrocket. Meanwhile, half
the humans on the planet—mothers, fathers, sisters, and brothers—
can't find sufficient food and water. A few are taking things into
their own hands, inciting conflict and violence. Perhaps they feel
they have no choice. Many of the rest of us do have choices. Future
generations will either lament our shortsighted refusal to face up
to the problems of our time or marvel at our creativity and courage.

The good news is that we have remedies—technologies, policies, and model programs. The authors in this book describe many
of them. But this phenomenal innovation won't move forward without greater public support. Quite simply, we need more power. We
don't yet have the political or economic muscle to implement many

of the remedies. Too many people are still unaware of any need for change, or they remain in denial. Too few feel hope. Others cling to simplistic descriptions of reality. The dominant power structure reinforces this inaction through extraordinary distraction from the real issues of our time. We are considered patriotic if we shop and anti-American or quixotic if we work for change.

Ultimately, the authors in this book are posing one central question: *What really matters?* This question illuminates their vision of a different future. Can we refocus our attention on this question as we make choices about our future?

There are useful parallels between our current dilemma and at least one earlier struggle. The issues at the heart of the nuclear power debate in the late 1970s were similar to the tensions currently underlying the quest for sustainability. Those in favor of nuclear technology promised a better quality of life, cheap energy, good jobs, and a clean environment. Energy for unlimited economic growth was to be available almost "cost-free," economically and environmentally. The risks were manageable and the waste products could be safely buried for thousands of years. Yet the highly consolidated nuclear industry and its utility allies failed to calculate or reveal the true costs of nuclear energy—in particular, the need to store and ultimately dispose of thousands of tons of highly radioactive waste for the next 250,000 years (the projected half-life of plutonium) as well as the potential health threats to nuclear workers and the general public. Advocates of safe energy were able to capitalize on the hidden costs of nuclear energy and through legal and political channels bring those costs to public attention. The result was to bring the nuclear expansion to a standstill.

Likewise, today's dominant model of economic expansion on a global scale hides the true costs of doing business. We are told that growth-oriented economies and corporate-dominated development are the path to poverty alleviation, vibrant democracies, and environmental stability. Increasingly, these claims ring hollow. Promises of universal prosperity tied to the immutable forces of economic globalization and ever-increasing consumption have lost credibility with millions of citizens around the globe.

Several important and relevant lessons stand out from the an-

tinuclear movement: (1) It took many years to raise public awareness and engage numerous key constituencies in the struggle for
change, but ultimately those public education and organizing
efforts were fundamental to success; (2) the critique of nuclear
energy by prestigious scientists and economists was crucial in influencing both public and elite opinion; (3) a national network of
elected officials, ratepayers, young activists, academics, health
professionals, labor union members (including nuclear plant workers), and artists, along with over one thousand community-based
groups, constructed broad-based coalitions that worked collabora-
tively for positive change; (4) nuclear power as a dominant economic model was stopped at the local and state levels, through public utility commissions, state legislatures, and the courts rather
than at the federal level, through policy reform; (5) people involved
in the struggle were able to articulate a clear vision of feasible alternatives that could safely replace nuclear energy as affordable
sources of electricity—decentralized and plentiful renewable energy, conservation, and a slow transition to a hydrogen economy;
(6) victory was connected to the marketplace—when nuclear energy became risky for investors and too costly for consumers, plans
to build new plants were canceled; (7) the network of organizations
and experts were sufficiently organized and connected to bring
nearly half a million people to Washington, D.C., just six weeks after
the Three Mile Island nuclear accident, in an outpouring of human
protest, adding to the pressure for change; and (8) the issue cap-
tured the imagination and sparked the involvement of countless
young idealists and activists who brought their boundless energy
to the struggle.

In the end, further development of nuclear power was stopped
in its tracks in the United States because the public began to
see it as a public-health threat. The risks of nuclear technologies
outweighed the potential benefits. When diverse constituencies
banded together to insist that the industry pay the true costs of
production, investors shied away and the industry could no longer
muster the capital for new plants. No new plants were ordered in
this country after 1978.

The accelerating pace of climate change linked with catastrophic shortages of food and clean water worldwide are proof that time is running out to create a sustainable planet. While avoiding the frenzy of survivalist panic, how can we create a political climate for sustainable economic policies, technologies, and household practices? What can we learn from our political victories in the past?

RAISE PUBLIC AWARENESS AND BUILD A LARGER BASE OF SUPPORT

There's no getting around it. People are the source of power. As citizens, workers, investors, consumers, and innovators, people make things happen. The overwhelming priority in the short term is to raise public awareness and attract more people to our cause. (Throughout this chapter, when I refer to "we," I mean a broad network of concerned individuals, nonprofit groups, companies, government agencies, students, academics, and civic leaders dedicated to sustainable development.) Some dismiss public education as too time-consuming and costly. Yet, without greater public support—not just reflected in opinion polls but manifested in human actions—few of the solutions outlined in this book will gain traction.

Why aren't more people getting involved? For some, it's simply lack of understanding. A large number of Americans have not connected the dots, often through no fault of their own. Many still don't see any relationship between driving a large vehicle and global-warming gas emissions or rising consumption of paper and the loss of ancient forests. We must help people make these kinds of associations.

Once individuals grasp the scope of the problem, they often resort to denial. Political leaders reinforce this resistance to change, proposing remedies that skirt the real problems at hand. This has been blatantly obvious with the debate over global warming where American political leaders have continuously deflected public attention from the urgent need to curb fossil fuel consumption while the problems associated with greenhouse gases rapidly worsen.

Ronald Heifetz, director of the Leadership Education Project at the John F. Kennedy School of Government at Harvard University, argues that people's defenses deserve respect. "People need time to see their lives in a different light—to change their images of the future." Those working to galvanize the public often feel we don't have time to help the public adapt. They must change now! Consequently, sustainability advocates come across as Chicken Little, shrieking that the sky is falling. Even if the facts are stark, this approach just isn't working. In fact, recent polls show a decline in public concern for both social and environmental issues.

Lack of awareness and denial of the problems are only two obstacles to engaging people. Extreme individualism is another. Many Americans draw the boundaries of their self-interest very narrowly. Our culture's emphasis on individualism and competition reinforces an attitude of isolation and impotence toward global problems.

Finally, while political leaders play to the public's desire for easy fixes, environmentalists err in the opposite direction, provoking undue distress by constantly describing ecological devastation and human suffering. People turn away from these overwhelmingly negative messages.

We clearly have not yet mastered the science and art of broadening and deepening public support for our cause. The private sector sells shoes, cars, potato chips—and introduces *some fifty thousand new consumer products in the United States each year*—by speaking to people's nonmaterial needs for love, friends, safety, and adventure. The goal is to make an emotional connection with people and then persuade them to take action, i.e., make a purchase. We must do a better job of speaking to the same nonmaterial needs—not to sell more stuff but to attract support based on the genuine human needs for hope, security, and greater connection. Too often, we overemphasize the bad news, prompting potential supporters to withdraw. What's more, individuals need to know that the actions they take will make an impact, and that they are not acting alone. These feelings of isolation or powerlessness are primary causes of burnout and dropout. To succeed, we must help people overcome denial, recognize their power to make a difference, overcome feelings of isolation, connect to deep values and aspirations, and focus on positive solutions more than negative consequences of inaction.

A CASE STUDY

I have spent much of the past decade wrestling with the thorny problem of U.S. consumption as a root cause of many environmental and social problems. As a nation, we consume a hugely disproportionate share of the Earth's resources and produce the largest volume of waste per capita. The environmental consequences of our "more is better" definition of the good life are severe, yet consumption is perceived as patriotic and even as the purpose for living.

I bring up consumption as an example of a problem that initially looks far too negative to take on and because my organization, the Center for a New American Dream, has made considerable progress in overcoming such communication challenges (though we still have much to learn). Our goal is to connect to people's deepest values and then support them in taking some positive action: to consume wisely (not to stop consuming) in households and workplaces or engage in national advocacy efforts. Our research indicated that people couldn't handle a lot of depressing data about the planet; they're too busy coping with day-to-day life, and therein was our entry point. Rather than focus first on the need to change consumption patterns to ensure environmental survival, we focused on the ways in which consumer culture feels bad to people in their daily lives—the sense that life is out of control and that the chase for more may not be worth the required effort. And rather than talk about having less, we talked about having more fun and fulfillment and focused on the benefits of resisting excess materialism. We stressed finding personal balance, simplifying one's life, and protecting kids from advertising. This positive approach to engagement has been well received. In short, we seek to help people move from denial to awareness to meaningful action.

EXPERIMENT WITH NEW APPROACHES TO PUBLIC OUTREACH

Data shows that Americans get insufficient sleep, work longer hours than their European counterparts, have no more than a week of annual vacation on average, and must juggle family and work with minimal support. It's difficult to engage people when they're tired or simply overwhelmed by all the bad news. In this context, how do we galvanize people to participate in the cracks of their otherwise very busy lives? Several organizations working to advance sustainability are achieving results by experimenting with new models for civic engagement.

The Appalachian Mountain Club builds trails, takes people on wilderness excursions, and subsequently signs up folks to push for policy reform to protect the lakes, mountains, and trees they've personally experienced. The South Carolina Conservation League in Charleston, the Chesapeake Bay Foundation in Annapolis, and the national Sierra Club similarly link environmental advocacy with personal experiences of nature. Each of these organizations tries to connect to the hearts of its members—not just to their wallets or computer terminals—by providing them with opportunities to experience the beauty of life, and then asking them to help protect the source of their wonder.

Alice Waters, owner and chef at Chez Panisse in Berkeley, California, follows a similar approach, helping people from all walks of life connect the dots from the vegetables on their plates to the farmers toiling in the fields. Her strategy: Link the pleasure of eating and the desire for good health with a willingness to take action on food production and consumption practices. From her small restaurant, she has literally launched a movement of children and adults dedicated to sustainable agriculture. Other organizations are exploring the successful strategies of groups such as Weight Watchers and Alcoholics Anonymous, which help people examine their addictions in the context of their genuine unmet needs for love, connection, and hope in a fractured world.

We can reach more people and establish longer-lasting relationships by connecting to their deeper aspirations. Activism is more likely to be sustained when people act because they love hiking in healthy forests, want to simplify their lives, yearn for more fun and less stuff, or feel a deep concern for malnourished children in southern Africa or Harlem.

CREATE NEW INSTRUMENTS FOR REACHING THE PUBLIC

For good or bad, most people receive information about the world primarily through electronic media, especially television. Americans in general spend 40 percent of their free time watching television and children are the most avid viewers, watching $2^3/_4$ hours on a typical day. Television programming and advertising is driven by commercial interests. Public debate is skewed and democracy is constrained because we have no alternative vehicle for reaching the public. Sustainability proponents need a noncommercial vehicle for news, feature stories, and family entertainment. Universities, public schools, nonprofit groups, and private companies should pool resources to purchase and launch a national cable channel devoted to commercial-free programming for a better world. Ambitious, but not impossible. This would revolutionize public debate and deepen public understanding about our world and where it's heading.

BUILD POWER AT THE LOCAL LEVEL

Satish Kumar, president of Schumacher College and a student of Mahatma Gandhi's movement for Indian independence, argues that one key to increasing our influence is to work first at the periphery of the dominant system before leveraging change at the center of power. Starting at the point of maximum resistance typically brings defeat. In other words, don't expect to go to Washington, D.C., or, in Gandhi's case, to the British Parliament to pass a new law or fundamentally revamp existing policy before building political support in communities, churches, unions, city councils, and state legislatures across the country. It took Gandhi fifty adult years to achieve success.

Many hoped for successful policy reform during the Clinton Administration, especially in support of sustainability, a new concept in the early 1990s. But when President Clinton proposed an energy tax early in his term, it was defeated within days, largely because no political infrastructure was in place to ensure success. Hopes for the elimination of harmful federal subsidies to mining, timber, and grazing interests, for new indicators of economic progress, a reordering of federal budget priorities, and for investment in urban revitalization and public transit were soon dashed. Within two years, the entire policy agenda for sustainable development had collapsed. There were many reasons for these defeats, but the main problem was a tendency to seek policy reforms in Washington without sufficient investment in building a political base to win the fight on Capitol Hill.

When it comes to energy policy and other model programs, we must continue to do what is politically possible in Congress, where the oil, gas, and automobile lobbies are formidable, while focusing more resources in local jurisdictions. We should put more emphasis in a place like San Francisco, where voters overwhelmingly approved two revenue bonds in November 2001 that will finance solar and wind energy systems, making the city the nation's largest municipal producer of solar electricity. Or in a place like Pennsylvania, where state colleges and universities are purchasing wind-generated electricity, making the state the largest purchaser of wind power in the United States.

Laws get passed by elected officials and we don't have enough enlightened leaders in office. Sustainability proponents need to identify and support local elected officials running as mayors, city council members, and state legislators. We need to build this power at the local level, keeping our attention on opportunities for wins at the Congressional level. If we hope to succeed, we need a cohesive team of elected leaders that collaborates across state boundaries and shares our vision of a sustainable world.

FOCUS ON REDIRECTING THE SYSTEM'S POWERFUL CURRENTS

The most powerful forces in American society today are enterprise and finance. The market economy is both the engine behind many social and environmental problems and potentially the mechanism for achieving positive change. It makes sense to use the power of market-based structures wherever possible. Think aikido, the Japanese martial art in which a physically weaker defender harnesses and uses the force of the more powerful opponent to his advantage.

In the past three decades, efforts to curb corporate excesses and promote responsible corporate behavior to protect the environment and worker rights have focused on the policy and regulatory arena. More recently, proponents of sustainability have shifted part of their efforts to the marketplace, and several recent initiatives suggest that this is precisely how to achieve greater short-term success. There are several ways to harness the power of the market to simultaneously promote economic prosperity, social development, and resource conservation.

Several successful market-based campaigns have integrated consumer demand, shareholder resolutions, student and nongovernmental organization (NGO) protests, and nonprofit-corporate partnerships to promote sustainability. The success of NGOs in changing forestry practices, working with progressive elements of the forestry industry, is now familiar to many people. The Forest Stewardship Council has certified over twenty million hectares of wood and many forests are being managed wisely as a result of consumer and environmental pressure. Likewise, environmental advocacy linked with strong consumer demand has led to increased sales of organic food, growing interest in the Marine Stewardship Council's certified seafood, and consumer rejection of genetically engineered food. More recently, the largest retail supplier of paper in the United States—Staples—responded to consumer and environmental demands with a pledge to offer a new line of paper with 90 percent post-consumer waste content and 10 percent tree-free fiber. Staples—under pressure from consumers and environmental groups—is in a position to influence the entire industrial infrastructure for recycled paper and help conserve forests.

We live in a market-based society, so when conservation organizations, in concert with consumers, investors, and selected producers, get organized, they have clout and often move more quickly than federal policy makers. Borders Books, Hyatt Hotels, Safeway, and Starbucks began selling coffee produced with sustainable growing methods in response to consumer and environmentalist demands. Dozens of coffees are now on the market, certified to protect the environment, workers, and wildlife.

Government agencies can also play a constructive role in the marketplace. Armed with annual budgets for goods and services totaling $385 billion, state and local governments alone represent one of the largest forces for developing the market for green products. Dozens of states, cities, and counties purchase paper with a high-recycled content, directly helping to save forests around the world. Santa Monica, California, is buying 100 percent renewable electricity for all of its city facilities, and numerous local agencies are buying nontoxic cleaning products to help improve workers' health and the local environment.

There are other ways to use the market for positive change. Large corporations have a global impact when they ask their suppliers to change certain practices. McDonald's, for example, with encouragement and pressure from environmental and animal rights groups, dramatically improved chicken production practices in the United States simply by specifying new production criteria for all its suppliers and insisting on more humane treatment of animals.

These initiatives take advantage of existing market forces and redirect consumer and investor dollars toward products and companies that demonstrate superior performance with regard to sustainability. Universities, local school systems, municipal governments, and faith-based institutions can make sustainable procurement and investment practices a priority. The shift in purchasing and investment power, in turn, can quickly alter the way goods are produced, improving human well-being and protecting the environment.

Businesses that promote sustainability are often rewarded in the marketplace. Their environmental and labor practices attract both investors and consumers while frequently reducing operational costs. When British Petroleum broke ranks to support the Kyoto treaty on global warming, the company changed the political atmosphere—and attracted the interest of socially responsible investors. When Goodman Manufacturing, a heating and cooling equipment company, protested the Bush Administration's plans to lower energy efficiency standards for household appliances, it helped keep the existing standards in place. In response, consumer and environmental groups generated several thousand letters from individual consumers promising to buy Goodman's appliances when they next made a purchase. Consumer and investor power can be used to reward companies that redesign products and production processes to promote sustainability.

CREATE THE NEW RATHER THAN ADAPT TO THE STATUS QUO

There is power in simply altering our relationship to the status quo. Author and activist Joanna Macy suggests that one path to gaining power involves stepping out of existing patterns of living by establishing new "self-organizing systems." Often we feel trapped, complicit with a system that is moving in a reckless direction. Macy's strategy involves altering the "givens" in our life. Imagine a circle dance where one, then two people move in a new direction. The others are inevitably influenced by the new pattern.

People need to see, touch, and experience the positive changes we advocate. This is why examples of lifestyles, communities, products, and businesses that promote sustainability are so powerful. Bill McDonough, one of the most visionary designers in the world, draws standing-room-only crowds, in part because he's a superb public speaker but also because he can show people a piece of fabric that is so safe, so nontoxic, that it is literally edible. Mark Ritchie and his Institute for Agriculture and Trade Policy in Minnesota are building new community models—farmers' markets linked with rural agricultural development—that make people feel good about both their personal health and their contribution to saving family farms. When we buy from local businesses, we create new interdependent relationships and more self-reliant communities.

People experience new possibilities when they drive a Toyota Prius and create new systems when they purchase fair-trade coffee. Citizens in Oakland, New Orleans, and the Bronx have altered their relationships with the status quo by replacing hazardous-waste sites with gardens, ball fields, and parks. And people begin to imagine new possibilities when they observe others who voluntarily consume less, work fewer hours, and play more.

New models give people hope in the face of all the bad news. When we see the beginnings of positive change, we know that more is not only possible but also probable. During the nuclear power debate, safe-energy festivals were popular. Parents and children would come to see and touch solar cookers, wind technologies, and inventive home designs for conservation, gaining a sense of excitement about the future. Today, Americans can enjoy the eleven thousand miles converted from rails to trails for biking and hiking and gain a vision of future transport and recreation. California's recent legislation forbidding the sale of sodas and junk food in schools helps parents recognize that schools should be commercial-free zones. Habitat for Humanity's straw-bale housing for low-income families demonstrates that sustainable development can and must address both a fractured environment and human needs.

Sometimes we just need to change one small piece in a larger interdependent system, and the ripple effect can be considerable. The good news is that this is happening. There are now school lunch programs supplied by local organic farmers, improving children's health and strengthening rural agriculture; car-sharing programs run by municipalities that strengthen community while reducing greenhouse gases; tool libraries that help families save money and reduce material consumption and waste. We see new models in government buildings made with native, sustainable materials; community banks with social and environmental lending criteria; micro-enterprise that integrates job creation and environmental restoration; publicly owned utilities and small businesses that generate electricity from renewable sources of energy while reducing electric bills; and cities that are converting parking lots back to paradise.

USE THE POWER OF NONVIOLENT ACTION

Sometimes, when other strategies fail, we must assert power by creating greater tension in the system. In his "Letter from Birmingham Jail," Martin Luther King described the importance of direct action. "Nonviolent direct action seeks to create such a crisis and foster such a tension that a community which has constantly refused to negotiate is forced to confront the issue. It seeks to dramatize the issue so that it can no longer be ignored." Sometimes this is the only way to challenge dominating power.

As I've already noted, antinuclear protesters sent a strong message to Congress when they mobilized nearly half a million Americans to come to Washington. This same strategy has been used most recently to force a discussion of globalization and trade. Unfettered capitalism and sustainable development are not compatible. There are obvious limits to economic growth and expansion on a finite planet. Our real task is to stop avoiding this debate and get on with establishing appropriate limits on enterprise while rewarding and providing incentives for private sector innovation that generates wealth and employment, conserves resources, and improves material security for people around the world. Direct action, led predominantly by students, has cracked open debate on this taboo subject. We must expose the hidden costs and obvious dysfunction of economic globalization and its mediating institutions while preparing for the inevitable negotiations over how best to restructure our global financial and trade policies. In the short term, direct action is a crucial strategy for fostering a healthy tension over these core questions.

More Unity, Less Chaos

Power usually comes through people working together toward common goals. It's remarkable how difficult this can be. Hyperindividualism and competition for resources often prevent key stakeholders from sharing information or collaborating. If we hope to succeed, we must work together. In 1992, at the end of the first Earth Summit, few organizations were addressing the root causes of escalating poverty and environmental degradation. Just a decade later, thousands of groups are taking on pieces of the sustainability puzzle—from technological innovation to policy reform—even as they collectively debate the true scope of the sustainability problem. Yet in the United States, these organizations, companies, and agencies have few mechanisms for assessing progress, reviewing objectives, sharing information, or building synergistic relationships. The few attempts at establishing national umbrella coalitions have failed.

This phenomenal growth in positive activity is encouraging, yet it poses challenges for getting work done, especially while everyone copes with the cascading wave of new information relevant to their work. The web of related organizations, experts, and activists grows exponentially each year. At a minimum, we need a few well-designed national and regional retreats that bring leaders together. During the nuclear power debates, annual national and regional conferences were held to exchange news and resources and forge common strategies. This was critical to holding the many constituencies and groups together.

We also need to help concerned citizens find easy access to our networks and groups. The movement to freeze nuclear weapons gained momentum in the 1980s by using a national petition, in concert with state referendums, to involve the public and build a political base of hundreds, and sometimes thousands, of citizen volunteers in every state. Something similar could be crafted to unify disparate elements of the sustainability community.

Finally, we must collaborate on the overarching messages we use to reach the public and on the specific actions we ask them to take. Too often, overlapping initiatives have a negative impact on the very people we need to enlist. Rather than providing the concerned public with a clear lighthouse—a beacon for finding their way—our collective impact is often experienced more as a cacophony of noise. The public simply turns us off. We need new incentives for cooperation. Leaders must build friendships, not just professional relationships, to help transcend the competition for resources and credit.

BEWARE OF PITFALLS

Don't Confuse Activity With Effective Action

We don't have time to get lost or distracted. Since the Rio Earth Summit in 1992, there have been scores of international and national papers, mandates, and conferences on various aspects of the summit's action agenda—Agenda 21—and sustainable development. This explosion of activity by stakeholder groups is overwhelmingly positive. The sharing of information has been and will continue to be crucial. Yet with so much going on, it is increasingly difficult to know what meetings to attend, which host organizations have follow-through capacity, and whether the bustle of activity has any political significance.

The utility of joint summits and statements depends on the priority the relevant stakeholders—governments, corporations, NGOs, and interested segments of civic society—give to the process and outcomes. Sometimes the flurry of activity over language substitutes for the more essential work of building public support for change. In the United States, for example, the federal government has largely ignored implementation of Agenda 21. International gatherings from Cairo to Johannesburg have absorbed tremendous human and financial capital from American NGOs with minimal impact on the U.S. government's behavior. International conventions are important but more effort should be focused on building domestic constituencies capable of influencing federal decision makers.

At the international level, the United Nations and other international agencies should seek to streamline the hundreds of organizations and forums. It is impossible for any stakeholder to monitor the UNEP, CSD, UNDP, FAO, UNESCO, OECD, and other institutions addressing sustainable development.

Be Conscious of Psychological Obstacles to Success

Most of us are familiar with political and economic obstacles to change: The pervasive influence of money in politics, corporate domination of the media, and inadequate financial resources for the cause, to name a few. But there are less obvious pitfalls and basic human tendencies that frequently slow things down. We often opt for simplistic analysis—describing reality in terms of good and evil and blaming authority figures or other scapegoats for all problems. We hold on to past assumptions and fail to adapt to changing realities. We tend to externalize the enemy and fail to examine our own values and behavior. We jump to conclusions and get distracted by power struggles. It is helpful to be conscious of these underlying barriers to progress as we leverage our increasing power to make positive change.

The Rising Threats of Centralized Financial Power and Militarism

Two trends make our work more difficult. The first is the weakening of democratic government in both the United States and many other countries. Multinational corporations and their major institutional stockholders have, to a large extent, replaced governments as the supreme power brokers on the world stage. The largest one hundred corporations have incomes greater than half the member countries of the U.N. *Five hundred corporations direct 70 percent of all international trade.* These corporations exert tremendous influence over elected officials and often undermine success in the federal policy arena. Political democracy, economic democracy, and environmental sustainability cannot be separated. There are many implications for strategy, but one stands out: Every constituency that cares about sustainable development should join the national effort to further reform campaign finance laws.

The second trend is militarism—as government has become weaker, the military has gained power. The connection between security issues and sustainability is clear. Resource scarcity coupled with widespread human suffering creates perfect conditions for conflict, terrorism, and war. The United States is responding to this situation by placing greater emphasis on military readiness and less on foreign aid or debt relief. There is growing concern about America's military intentions, especially U.S. willingness to protect access to foreign oil at any cost. Similarly, many people are disturbed by rising American nationalism in the wake of the World Trade Center attacks on September 11, 2001.

As this book goes to press, the Bush Administration has proposed the largest military spending increase in two decades and is rapidly seeking to modernize nuclear weapons. The Anti-Ballistic Missile Treaty has been abandoned and the Nuclear Weapons Council, made up of officials from the U.S. Defense and Energy departments, has ordered a three-year study of the development of a nuclear-tipped, earth-penetrating weapon that can destroy hardened underground targets. Opposition to these policies has been weak and muted.

These trends, in concert with growing concern about the loss of privacy, civil liberties, and a free and fair press, must be monitored carefully. Ultimately, sustainable development depends on a robust civic democracy. Strategies for change must incorporate partnerships with those working for campaign finance reform, a reduced role for the military in our society, and increased public ownership of and access to communications channels.

FROM WHENCE COMES OUR POWER?

As we make our way in a wounded world we need to do our best to stay connected to what is real and good. So much is artificial and superficial. Albert Einstein was once asked, "What is the most important question you can ask in life?" His response: "Is the universe a friendly place or not?" Most of us don't feel too certain of the answer, but Einstein believed the evidence came down in favor of goodness, that the universe is ultimately friendly. Whether you consider the universe friendly, neutral, or hostile is a matter of faith, but it's important to have personal strategies for tapping back into some source of strength beyond the self. Otherwise, it's too easy to get preoccupied with protecting our precarious isolated existence, fighting depression and a sense of isolation.

From whence comes our power? Ultimately, to stay grounded in a world that is out of whack, we must deepen our connections—to the natural world, to other people, and to the sacred, however we define that. We tend to think genuine power comes from exertion, smarter strategies, and more resources. These things can help on a tactical level, but our sense of vision and creative courage come more from surrender than resistance, from letting go than holding on, from trusting in the friendliness of the universe despite a preponderance of fear.

Proponents of sustainable development frequently burn out, have health problems, and lose our way. The demands of work frequently squeeze out other life priorities, including practices that keep us in balance. Some of us even get addicted to the intensity and pressure. Everything seems urgent. We must rest more and do less if we hope to stay the course. We need to experience more gratitude and less fear. It's not an intellectual or analytical thing. It's the letting go of resistance, of having to know what to do, even the letting go of results once we've given our best, and just resting in the beauty of what is. Gratitude comes from the opposite of struggle. It comes from acceptance of the good in life, from the belief that the universe tilts toward justice, peace, and human understanding.

When we take time to watch a sunset, meet a friend, or seek truth in silence, we tap back into those connections that sustain us through this work. If we hope to contribute to a better world, we cannot afford to operate strictly on our own juices. That puts a tremendous burden on each individual—an overwhelming need to compete, defend, and construct strategies for survival. If we go it alone, we must always ask, "How do I do this?" If we stay connected, we can spend more time listening, paying attention, and relaxing into this journey. Less effort sometimes leads to extraordinary outcomes.

All the great spiritual traditions suggest that authentic moral action stems from opening to the possibility of higher forces beyond the individual self. And while our struggle includes courageous political action, it is equally a struggle of our inner hearts. Perhaps our greatest need in this time of global and personal disequilibrium is for reconnection to the possibility of inexhaustible love. The quest for sustainable development is really the quest for more love in the world—for care of the natural world, of innocent people, and future generations. Our work is really about giving love back to the world and we can't give what we don't have. Try trusting that love is in our genetic code, that we are meant for it, and that it is our ultimate source of wisdom. It involves surrendering to not knowing while hungering to know. And perhaps this reconnection to what matters—through nature, human relationships, and the sacred dimension of life—will be the true source of power that takes us from here to there.

NOTES

MCDONOUGH/BRAUNGART: THE EXTRAVAGANT GESTURE

15   *A new report from the World Resources Institute:* Emily Matthews, *The Weight of Nations: Material Outflows from Industrial Economies* (Washington, DC: World Resources Institute, 2000).

18   *doing things right:* Peter Drucker, *The Effective Executive* (New York: HarperBusiness, 1986).

22   *a significantly higher degree of job satisfaction:* Judith Heerwagen, "Do Green Buildings Enhance the Well Being of Workers? Yes," *Environmental Design and Construction,* July–August 2000.

26   *"what was many becomes one":* Herman Daly, "Globalization Versus Internationalization: Some Implications" (Talk delivered in Buenos Aires, November 1998).

27   *Not the members of Slow Food:* Alexander Stille, "Slow Food," *The Nation,* August 20, 2001.

SCHOR: CLEANING THE CLOSET

See: www.newdream.org/publications/schornotes.html

HOLLENDER: CHANGING THE NATURE OF COMMERCE

65   *the fourteen warmest years on record in just the last twenty-two:* Denis Hayes, *The Official Earth Day Guide to Planet Repair* (Washington, DC: Island Press, 2000), 10–11.

65   *20 percent of the world's forest cover:* Navin Ramankutty and Terry Devitt. "Is Land Use Even More Important than Global Warming?" *Daily University Science News,* July 12, 2001.

65   *Estimated extinction rates:* William K. Stevens, "How Many Species Are Being Lost?" *New York Times,* July 25, 1995, C4.

65   *Synthetic organic chemical production:* Joe Thornton, *Pandora's Poison* (Cambridge, MA: MIT Press, 2000).

66   *ZERI founder Gunther Pauli quote:* "ZERI: A Youngster with Influence," *Chemical and Engineering News,* The American Chemical Society, July 8, 1996.

66   *Since 3M officially acknowledged:* Businesses for Social Responsibility On-line Resource Library, Topic Overview: Water Quality, www.bsr.org/BSRLibrary/TOdetail.cfm?DocumentID=529.

69   *the Pennsylvania Power and Lighting Company:* A. K. Townsend, *The Smart Office* (Olney, MD: Gila Press, 1997).

70   *pollution output equal to six million cars:* Ibid.

71   *job losses due to environmental regulations:* Michael Renner, *Working for the Environment: A Growing Source of Jobs* (The Worldwatch Institute, September 2000).

71   *states with the best environmental records:* Keith Ernst, Chris Kromm, and Jaffer Battica. *Gold and Green 2000* (The Institute for Southern Studies, November 2000).

75   *organic foods . . . industry sales are growing at nearly 23 percent per year:* The Natural Marketing Institute and the Organic Trade Association, Organic Consumer Trends 2001, www.nmisolutions.com.

RITCHIE: BE A LOCAL HERO

95   *such as the gigantic hog factories:* For example, see Attracting Consumers with Locally Grown Products, Food Processing Center, University of Nebraska-Lincoln, December 2001, available at www.farmprofitability.org.

96   *added value to each other's products:* Thomas Petzinger, Jr., *The New Pioneers: The Men and Women Who Are Transforming the Workplace and Marketplace* (New York: Simon and Schuster, 1999).

103   *"Kerala suggests a way out":* Bill McKibben, "The Enigma of Kerala," *Utne Reader,* March 1996, p. 111.

103   *"This is the direct opposite":* Ibid., p. 111.

104   *"Kerala does not tell us":* Ibid., p. 112.

105   *a beautiful and inspirational book: Renewing the Countryside* (Minneapolis, MN: Institute for Agriculture and Trade Policy, 2001).

DITTMAR: SPRAWL

109   *In 1998, there were 184,980,187:* Bureau of Transportation Statistics, Transportation Statistics 2001, p. 61.

109   *In 1995, 91 percent:* Ibid., p. 61.

109   *The wealthier a household is:* Patricia S. Hu and Jennifer R. Young, *Summary of Travel Trends: 1995 Nationwide Personal Transportation Survey,* Federal Highway Administration, 1997, p. 43.

109   *In 2000, Ad Age magazine estimates:* "100 Leading National Advertisers," Ad Age, 2001, www.adage.com/page.cms?pageId=639.

110   *During the same period, driving grew:* Bureau of Transportation Statistics, Transportation Statistics 2001, p. 45.

110   *Thus, the average household stayed roughly the same size:* Federal Highway Administration, 1995 Nationwide Personal Transportation Survey, p. 71.

110   *Catherine Ross and Anne Dunning examined:* Catherine L. Ross and Anne Dunning, "Land Use Transportation Interaction: An Examination

of the 1995 NPTS Data," prepared for the Federal Highway Administration, 1997, p. 14.

111 *More than one-third of U.S. carbon dioxide emissions:* F. Kaid Benfield, Matthew D. Raimi, and Donald D.T. Chen, "Once There Were Greenfields," Natural Resources Defense Council, March 1999, p. 49.

111 *At the global level, Americans consume over one-third:* Ibid., p. 50.

111 *Over 41,000 Americans die each year on highways:* Bureau of Transportation Statistics, Transportation Statistics 2001, p. 99.

112 *A recent National Research Council report:* Transportation Research Board, Strategic Highway Research, Special Report 260, 2001.

112 *The American Automobile Association estimated:* Bureau of Transportation Statistics, Transportation Statistics 2001, p. 204.

113 *Nationally, transportation expenditures account for 17.5 percent:* "Driven to Spend," Surface Transportation Policy Project and Center for Neighborhood Technology, www.transact.org, 2000, p. 3.

113 *The poorest quintile of American households:* Ibid., p. 3.

113 *Scott Bernstein and Ryan Mooney of the Center for Neighborhood Technology:* Scott Bernstein and Ryan Tracey Mooney, "Driven to Debt: When the American Dream Prevents the American Dream," forthcoming from the Brookings Institution on Urban and Metropolitan Policy, 2003, p. 3.

113 *This study, which analyzed odometer readings:* John Holtzclaw et. al., forthcoming in *Transportation Planning and Technology,* 2002.

115 *Ten thousand dollars invested in a car declines:* "Driven to Spend," Surface Transportation Policy Project, www.transact.org, 2000, p. 6.

116 *The rate of growth in driving has dropped:* "Americans Flock to Transit,

Ease up on Gas Pedal" Surface Transportation Policy Project, 2001, www.transact.org/Pressroom/vmt-transit.htm.

116 *Transit use grew by 11 percent:* Ibid.

117 *A recent study by Rebecca Sohmer and Robert Lang:* Rebecca A. Sohmer and Robert E. Lang, *Downtown Rebound,* Fannie Mae Foundation and Brookings Institution Center on Urban and Metropolitan Policy Census Note, May 2001, p. 6.

118 *A recent study by Jones Lang LaSalle:* Jones Lang LaSalle, Property Futures, 2001, www.joneslanglasalle.com.

118 *The annual Emerging Trends in Real Estate report:* Real Estate Research Corporation, *Emerging Trends in Real Estate.* 1999, www.rerc.com/historicalemergingtrends.html.

119 *One developer in the Portland, Oregon, area put it this way:* "Transit-Friendly Housing," *Builder,* National Association of Home Builders, 1998.

119 *Car-sharing programs were being introduced:* Kenneth Orski, Innovation Briefs, 2002.

120 *William Frey projects:* William Frey, "Melting Pot Suburbs: A Census 2000 Study of Suburban Diversity," Brookings Center on Urban and Metropolitan Policy, June 2001.

120 *According to Catherine Ross and Anne Dunning's analysis:* Catherine L. Ross and Anne Dunning, "Land Use Transportation Interaction: An Examination of the 1995 NPTS Data," prepared for the Federal Highway Administration, 1997, p. 21.

121 *A recent study found that 57 percent of this generation:* "Moving Ahead: The American Public Speaks about Roadways and their Communities," Federal Highway Administration, 2001.

121　*The 2000 census found that nonfamily households:* Tavia Simmons and
Grace O'Neill, "Households and Families: 2000," Census Brief, U.S.
Census Bureau, September 2001, pp. 1-2.

122　*Dowell Myers at the University of Southern California estimated:* "The
Coming Demand," Congress for the New Urbanism, 2001, pp. 4-6.

PARTHASARATHI: TOWARD PROPERTY AS SHARE

This essay relies on a large number of bibliographic sources. Readers
interested in exploring them can consult the following: For classic
accounts of property rights in Europe, see R. H. Tawney, *The Agrarian
Problem in the Sixteenth Century* (London: Longmans, Green and Co.,
1912); J. L. Hammond and Barbara Hammond, *The Village Labourer 1760-
1832: A Study in the Government of England before the Reform Bill* (London: Longmans, Green and Co., 1911); Marc Bloch, *French Rural History: An
Essay on its Basic Characteristics,* trans. Janet Sondheimer (Berkeley: University of California Press, 1970).

For more recent explorations, see John Rule, "The Property of Skill in
the Period of Manufacture," *The Historical Meanings of Work,* ed. Patrick
Joyce, (Cambridge: Cambridge University Press, 1989); J. M. Neeson, *Com-
moners: Common Right, Enclosure and Social Change in England, 1700-
1820* (Cambridge: Cambridge University Press, 1993); and Jan Luiten van
Zanden, "The Paradox of the Marks. The Exploitation of Commons in the
Eastern Netherlands, 1250-1850." *The Agricultural History Review,* vol. 47
(1999), pp. 125-44.

For further information on property rights in South Indian agriculture,
see Christopher John Baker, *An Indian Rural Economy, 1880-1955: The
Tamilnad Countryside* (Delhi: Oxford University Press, 1984); David Lud-
den, *Peasant History in South India* (Princeton: Princeton University Press,
1985); S. S. Sivakumar and Chitra Sivakumar, *Peasants and Nabobs: Agrarian Radicalism in Late Eighteenth Century Tamil Country* (Delhi: Hindustan
Publishing Corporation, 1993); Prasannan Parthasarathi, *The Transition to
a Colonial Economy: Weavers, Merchants and Kings in South India, 1720-
1800* (Cambridge: Cambridge University Press, 2001); Prasannan Parthasarathi, "State, Community and Property in the 18th Century: A Compara-

tive Perspective," Paper Presented at the Global History Workshop, International Institute of Social History, Amsterdam, November 2000.

An extensive discussion on institutions for the successful use of common property may be found in Elinor Ostrom, *Governing the Commons: The Evolution of Institutions for Collective Action* (Cambridge: Cambridge University Press, 1990).

ANDERSON/CAVANAGH: ANOTHER WORLD IS POSSIBLE

156 *gains in economic growth:* "Is Growth Enough?" Inter-American Development Bank Research Department, vol. 14, 2001, p. 3.

157 *United Nations studied seventy-seven countries:* Human Development Report 2001, United Nations Development Program, p. 17.

157 *sharp decline from an annual per capita growth:* Mark Weisbrot, Dean Baker, Egor Kraey, and Judy Chen, "The Scorecard on Globalization 1980-2000: Twenty Years of Diminished Progress," Center for Economic and Policy Research, July 9, 2001, p. 2.

157 *from $586.7 billion to more than:* Global Development Finance 2001 World Bank, (Washington, DC: World Bank, 2000), p. 246.

158 *from $4.4 billion in 1993:* World Bank, *Global Development Finance 2001* (Washington, DC: World Bank, 2000), p. 380.

158 *the country's exports to the United States:* U.S. Census Bureau, U.S. Trade Balance with Mexico, www.census.gov.

158 *percentage of Mexicans living in poverty:* "Memorandum of the President of the IBRD and IFC to the Executive Directors on a Country Assistance Strategy Progress Report of the World Bank Group for the United Mexican States," World Bank, May 21, 2001, p. 4.

159 *pollution from Mexican manufacturers nearly doubled:* Kevin P. Gallagher, "G-DAE Working Paper No. 00-07: Trade Liberalization and

Industrial Pollution in Mexico: Lessons for the FTAA," Global Development and Environment Institute, Tufts Univerity, October 2000, p. 14.

159 *The value of the median wage:* "Paycheck Economics," Economic Policy Institute, www.epinet.org/paycheck (2000).

159 *executive pay jumped 571 percent:* Sarah Anderson et al. "Executive Excess 2001," Institute for Policy Studies and United for a Fair Economy, 28 September 2001, p. 3.

159 *In a study of hundreds of union organizing campaigns:* Kate Bronfenbrenner, "Uneasy Terrain: The impact of capital mobility on workers, wages and union organizing." Commissioned Research paper for the U.S. Trade Deficit Review Commission, September 6, 2000, p. 20.

161 *These views are summed up in a document:* This document is available in English and Spanish on the following website: www.asc-hsa.org.

162 *a total debt burden of $809 billion:* World Bank, *Global Development Finance 2001,* (Washington, DC: World Bank, 2000), p. 252.

166 *the IFG asserts that healthy societies:* This section is adapted from International Forum on Globalization "Alternatives" Task Force, *A Better World Is Possible: Alternatives to Economic Globalization* (San Francisco: International Forum on Globalization, 2002).

RECHTSCHAFFEN: TIMESHIFTING

177 *Instead, between 1977 and 1997, the average workweek:* James Lardner, "World-Class Workaholics," *U.S. News & Word Report,* December 20, 1999.

178 *We could now reproduce our 1948 standard of living:* Juliet B. Schor, *The Overworked American: The Unexpected Decline of Leisure* (New York: Basic Books, 1991), p. 2.

178 *Instead, in the 1990s alone, Americans increased the hours worked:* Ste-

ven Greenhouse, "For Americans, a Day at the Office Keeps Growing," *New York Times,* September 4, 2001.

178   *In 2000, the average employed American worked three and a half more weeks.:* Ibid.

178   *U.S. workers average a paltry thirteen vacation days a year:* "International Travel Data," Travel Industry Association of America, www.tia.org/press/itd.asp#vacation.

178   *Credit card debt per U.S. household was $8,488 in mid-2001:* "Card Debt," CardData.com, statistical data bank of U.S. credit card industry, www.cardweb.com/cardtrak/news/2001/august/31a.html.

185   *If I am incapable of washing dishes joyfully:* Thich Nhat Hanh, *Peace Is Every Step: The Path of Mindfulness in Everyday Life* (New York: Bantam, 1992), p. 26.

186   *annual working hours of the average U.S. employees have increased 180 hours:* Larry Mishel, Jared Bernstein, and John Schmitt *State of Working America,* 2000/1, (Ithaca, NY: Cornell University Press, 2001), Table 2.1, p. 115.

186   *Working hours for parents, author's calculations from:* Ibid., Table 1.29, p. 98.

187   *The value of the minimum wage has eroded:* Ibid., Table 2.40, p. 1.

TAYLOR: HOW DO WE GET FROM HERE TO THERE?

233   *the Brundtland Commission defined sustainable development:* Our Common Future, World Commission on Environment and Development, 1987.

236   *"People need time to see their lives in a different light":* Ronald Heifetz, *Leadership without Easy Answers* (Cambridge, MA: Harvard University Press, 1994).

237 *recent polls show a decline in public concern:* www.GreenbergRe-
search.com.

237 *some fifty thousand new consumer products in the United States each
year:* From www.neweconomyindex.org/section1_page09.html.

240 *Americans in general spend 40 percent of their free time watching television:* Author's calculations from John P. Robinson and Geoffrey God-
bey, *Time for Life* (University Park, PA: Penn State Press, 1997), Chs.
8 and 9.

240 *Children watch $2^3/_4$ hours of television:* Kids & Media@The Millenium,
Kaiser Family Foundation, available at kff.org, 1999, Table 8A.

245 *Letter from Birmingham Jail:* almaz.com/nobel/peace/MLK-jail.html.

249 *Five hundred corporations direct 70 percent of all international trade:*
"Ending Corporate Governance," Richard Grossman in Dean Ritz, ed.,
Defying Corporations, Defining Democracy (New York: The Apex
Press, 2001), p. 23.

CONTRIBUTORS

Sarah Anderson is the director of the Global Economy Program at the Institute for Policy Studies in Washington, D.C. Anderson is the coauthor (with John Cavanagh and Thea Lee) of the book *Field Guide to the Global Economy* and a board member of the Alliance for Responsible Trade and the Coalition for Justice in the Maquiladoras.

Michael Braungart is a chemist and the founder of the Environmental Protection Encouragement Agency, a scientific consultancy in Hamburg, Germany. In 1995 he cofounded McDonough Braungart Design Chemistry, a product and system development firm assisting companies such as Ford Motor Company, Nike, Herman Miller, and BASF in implementing their unique sustaining design protocol.

John Cavanagh is director of the Institute for Policy Studies and coauthor of ten books on the global economy, most recently *Field Guide to the Global Economy* (with Sarah Anderson and Thea Lee). He is chair of the Alternatives Task Force of the International Forum on Globalization and has previously worked at the World Health Organization and the United Nations Conference on Trade and Development.

Herman E. Daly is professor at the University of Maryland School of Public Affairs, formerly a senior economist at the World Bank, and author of *Steady-state Economics* and *Beyond Growth.*

Hank Dittmar is president of the Great American Station Foundation, a national organization dedicated to the restoration of historic rail stations and the communities surrounding them. He is a national authority on transportation policy management and has served variously as an advocate for more sustainable transportation systems, a regional planner, an airport director, and a transit manager.

Robert Engelman is vice president for research at Population Action International, a Washington-based research and advocacy organization. A former environmental journalist, he serves as a visiting lecturer at Yale University and chairs the board of the Center for a New American Dream.

Peter Forbes is a conservationist, writer, and photographer who lives and farms in the Mad River Valley of Vermont. He is a vice president of the Trust for Public Land and the director of the Center for Land and People. He is the editor of *Our Land, Ourselves* and is the author of *The Great Remembering: Further Thoughts on Land, Soul and Society.* Peter is on the board of directors of the Center for a New American Dream. You can reach him at Peter.Forbes@TPL.org.

Jeffrey Hollender is the president and CEO of Seventh Generation, a thirteen-year-old company based in Burlington, Vermont. Seventh Generation is the leading U.S. marketer of safe and environmentally friendly household products. Jeffrey's book, *How to Make the World a Better Place: A Guide for Doing Good,* was originally published in January of 1990 and a revised edition was released in March of 1995.

William McDonough is an architect, industrial designer, and educator. He is a founding principal of William McDonough & Partners, Architecture and Community Design, and McDonough Braungart Design Chemistry, both based in Charlottesville, Virginia. McDonough and Braungart's new book on the transformation of human industry, *Cradle to Cradle: Remaking the Way We Make Things,* was published in 2002.

Bill McKibben is the author of *The End of Nature, The Age of Missing Information,* and a forthcoming book on human genetic engineering, *Enough.* He is currently a visiting scholar at Middlebury College.

Prasannan Parthasarathi received a doctorate in economics from Harvard University and is an associate professor of history at Boston College. He is the author of *The Transition to a Colonial Economy: Weavers, Merchants and Kings in South India, 1720–1800.*

Dr. Mary Pipher is a clinical psychologist and author of many books, including *The Shelter of Each Other* and *The Middle Of Everywhere: The World's Refugees Come to Our Town.* She is on the board of the Center for a New American Dream.

Stephan Rechtschaffen, MD is the author of *Timeshifting: Creating More Time to Enjoy Life* and is the cofounder and CEO of Omega Institute, the largest holistic educational center in the country. He is a pioneering holistic physician who has been exploring alternative and complementary approaches to health and well-being.

Mark Ritchie is president of the Institute for Agriculture and Trade Policy and a lifelong advocate for local food systems, family farming, and sustainable communities. He can be reached at mritchie@iatp.org

Vicki Robin is the coauthor with Joe Dominguez of the national best-seller *Your Money or Your Life: Transforming Your Relationship With Money and Achieving Financial Independence,* which is now available in nine languages. Vicki is president of the Seattle-based New Road Map Foundation, committed to teaching people tools for sustainable living. She is chair of the Simplicity Forum and founder of Conversation Cafés, fostering a culture of conversation through drop-in dialogue in public spaces.

Juliet B. Schor is professor of sociology at Boston College and the author of *The Overworked American: The Unexpected Decline of Leisure* and *The Overspent American: Why We Want What We Don't Need.* She is also a founding board member of the Center for a New American Dream.

Betsy Taylor is the founder and executive director of the Center for a New American Dream. She previously served as executive director of the Merck Family Fund, vice chair of the Environmental Grantmakers Association, and as a member of the Population and Consumption Taskforce of the President's Council for Sustainable Development. She holds a master's degree in public administration from Harvard University and a B. A. from Duke University.

Congresswoman **Nydia M. Velázquez** is the first Puerto Rican woman elected to the United States House of Representatives. Congresswoman Velázquez was the first person in her family to receive a college diploma. She received a bachelor's degree in political science from the University of Puerto Rico and a master's in political science from New York University.

In February of 1998, Congresswoman Velázquez was named ranking Democrat on the House Small Business Committee, making her the first Hispanic woman to serve as chair or ranking member of a full committee in the history of the House of Representatives. She serves the Twelfth District of New York, which encompasses parts of Brooklyn, Manhattan, and Queens.

RESOURCES

For more information about some of our authors and their organizations and affiliations, please see below.

Omega Institute for Holistic Studies
150 Lake Drive
Rhinebeck, NY 12572
800-944-1001
845-266-4444
fax: 845-266-3769
www.eomega.org

AWAKENING BODY, MIND AND SPIRIT

Omega was founded in 1977 at a time when holistic health, psychological inquiry, world music and art, meditation, and new forms of spiritual practice were just budding in American culture.

Since then, Omega has become the nation's largest holistic learning center. Every year more than twenty-thousand people attend workshops, retreats, and conferences on our 140-acre campus in the countryside of Rhinebeck, New York, and at other sites around the world.

While Omega has grown in size, their mission remains the same. Their programs feature all of the world's wisdom traditions. They are not aligned with any particular healing method or spiritual tradition. The world changes one person at a time. Therefore, they dedicate their work to the transformation of society, so that one day all beings will know health, happiness, and peace, and the world will live in balance and harmony.

Institute for Agriculture and Trade Policy
Mark Ritchie, president
2105 First Avenue South
Minneapolis, MN 55404
612-870-0453

fax: 612-870-4846
www.iatp.org
iatp@iatp.org

Founded in 1986, the Institute for Agriculture and Trade Policy (IATP) educates and assists individuals and groups working for a just and sustainable world. IATP's mission is to promote resilient family farms, rural communities, and ecosystems around the world through research and education, science and technology, and advocacy. They work toward local, state, national, and international policies that will shape a more equitable and democratic world economy; promote socially and ecologically sustainable development; ensure environmental protection, biodiversity, and food security; value human rights; and strengthen local and regional economies. Their goal is to make the relevant policy-making institutions understandable, accessible, and accountable, and to enable citizens to participate effectively in domestic and international decisions that affect their communities and regions.

IATP has a long-term commitment to policy innovation and advocacy on food, agriculture, trade, and environmental issues. They work with farmers, consumers, unions, environmental organizations, citizens' groups, and others—both in the United States and around the world—to influence the direction of policy making, including trade and investment agreements. They are working to create equitable trading relationships among different countries by putting forward positive fair-trade alternatives and by organizing direct sales between farmers and consumers, both within countries and across borders.

Population Action International
1300 19th Street NW
Second Floor
Washington, D.C. 20036
202-557-3400
fax: 202-728-4177
www.populationaction.org
pai@popact.org

Population Action International is a research and advocacy organization that builds support worldwide for population programs based on individual rights. At the heart of PAI's mission is its commitment to universal access to family planning and related health services, and to educational and economic opportunities for girls and women.

McDonough Braungart Design Chemistry
401 East Market Street
Charlottesvile, VA 22902
434-295-1111
fax: 434-295-1500
www.mbdc.com
info@mbdc.com

William McDonough & Partners
Architecture & Community Design
410 East Water Street
Charlottesville, VA 22902
434-979-1111
fax: 434-979-1112
www.mcdonough.com

CENTER FOR A NEW AMERICAN DREAM

The Center for a New American Dream helps Americans consume responsibly to protect the environment, enhance quality of life, and promote social justice. The center works with individuals, institutions, communities, and businesses to conserve natural resources, counter the commercialization of our culture, and promote positive changes in the way goods are produced and consumed.

Founded in 1997, the center has emerged as a leading voice in poviding hopeful, useful alternatives to the "more is better" definition of the American dream. Center campaigns have been featured in thousands of stories spread throughout virtually every major media outlet in the country, and center programs are making a difference in the way Americans consume resources. The center's Turn the Tide and Step by Step programs make it easy for individuals to take effective environmental and political action in their daily lives. The Procurement Strategies program helps government and other institutional purchasers who spend over $400 billion each year, to buy environmentally preferable products. The Youth and Faith-Based programs help these two very influential constituencies practice and promote sustainable consumption.

Resources from the Center for a New American Dream include books, brochures, an engaging video hosted by Danny Glover, and a very popular website. Supporting members of the center receive a More Fun, Less Stuff kit with tips for consuming wisely and enjoying the nonmaterial pleasures in life. For more information on the center or to become a supporting member, contact

Center for a New American Dream
6930 Carroll Avenue, Suite 900
Takoma Park, MD 20912
1-877-68-DREAM
www.newdream.org
newdream@newdream.org

ACKNOWLEDGMENTS

This book would not have been possible without the help of a number of individuals. For his careful editing and help in the early stages of the book, we would like to thank Karl Steyaert; for her tireless work in putting the manuscript together, communicating with authors, and cheering everyone on, we are grateful to Amy Rutledge; for his valuable input, we thank Eric Brown. We are also indebted to the staff and board of the Center for a New American Dream, without whom this project would not have been possible.

Juliet Schor would like to thank Larry Raff and Prasannan Parthasarathi, who contributed to her chapter. Betsy Taylor would like to thank Denny May, her toughest and most helpful critic, Amy Rutledge, and Michal Keeley, for thoughtful editing, and Peter Barnes, for generously providing a perfect place to write at the Mesa Refuge in Pt. Reyes, California. In addition, Hank Dittmar wants to thank Scott Bernstein, Sarah Campbell, John Holtzclaw, and Don Chen for their insights and advice. Jeffrey Hollender would like to acknowledge Geoff Davis for his extensive assistance with his chapter; without his help he could never have completed it. Mark Ritchie wishes to acknowledge the contributions of Dave Gutknecht, Kristen Corselius, Char Greenwald, and Derek Masselink to the writing of his chapter, and all of the hardworking people who are making these local food systems work. Vicki Robin wishes to thank Joe Dominguez (1938–1997), her great friend and teacher on the subject of money—and much else. Congresswoman Velázquez wishes to acknowledge Eric Brown for his indispensable assistance with her chapter. William McDonough and Michael Braungart would like to thank Chris Reiter for editorial assistance.

We are grateful to our editor at Beacon Press, Amy Caldwell, for her development of this book. Finally, we wish to acknowledge the many foundations and donors who support the Center for a New American Dream. Without their generous support, this book would not have been possible.

273